REPAIR AND REHABILITATION
OF
STEEL BRIDGES

REPAIR AND REHABILITATION OF STEEL BRIDGES

UTPAL K GHOSH

BE, CEng, FIE, MICE, MIStruct E, PE
CONSULTING ENGINEER

A.A. BALKEMA / ROTTERDAM / BROOKFIELD / 2000

A.A. Balkema, P.O. Box 1675, 3000 BR Rotterdam, Netherlands
Fax: +31.10.4135947; E-mail: balkema@balkema.nl
Internet site: http://www.balkema.nl

Distributed in USA and Canada by
A.A. Balkema Publishers, Old Post Road, Brookfield, VT 05036-9704, USA
Fax: 802.276.3837; E-mail: Info@ashgate.com

ISBN 90 5809 207 0

To
my wife, Manjula
and
our son, Indranil

Preface

A large number of steel bridges around the world were constructed during early 1900s and are now 75–100 years old. Many of these bridges may be showing signs of deterioration due to lack of maintenance or simply due to ageing. Some may be functionally obsolete to carry today's traffic. In such a situation the first suggestion naturally would be to replace these old bridges. This would mean investment of a huge amount of money. However, funds are scarce globally these days. It is being increasingly realised all over the world that the investment required to replace all these old bridges will not be available soon enough and that, in the mean time, maximum benefit should be derived from the limited available funds by rehabilitating these existing bridges to suit present day needs at a much lower expense. In other words, the strategy should be to focus our attention on the maintenance and management of the existing assets and make them serviceable wherever possible. In the backdrop of this scenario, rehabilitation and maintenance of the existing steel bridges have assumed particular significance. Thankfully, there is already a growing interest on the topic as is evident from the discourse in seminars and workshops being organised by different learned institutions as also from the technical papers being published currently.

Rehabilitation of steel bridges requires a high degree of ingenuity and expertise. It is true that steel enjoys an inherent advantage over other structural materials, because of its unique physical and chemical properties as also for its adaptability to various forms of connections. The process of arriving at a simple and practicable remedial solution, however, is far from simple. This is primarily because of the uniqueness of the character of almost every bridge, which is influenced by its location, environment, loading history, type of the structure, workmanship, quality of maintenance and many other factors. As a result it will not be appropriate to apply a uniform remedial method for repair and rehabilitation of all apparently similar types of bridges. Deficiencies of each bridge need to be studied individually, for arriving at a workable rehabilitation scheme.

However, not many books are available on the topic to meet the increasing demand. The primary aim of this book is to bridge this gap and provide the reader with an overall perspective on the subject. It attempts to highlight both the theoretical as well as the practical aspects which govern rehabilitation of steel bridges.

Many of the ideas presented in this book have been drawn from the author's own experiences and personal notes. However, there are also many areas where the author has taken the liberty of deriving basic information and ideas from diverse sources. As far as possible references to the published literature have been mentioned in the Bibliography at the end of each chapter. The author thankfully acknowledges his indebtedness of these earlier writers. However, it may well be that some of the ideas of earlier writers have appeared in the book without appropriate acknowledgement. If so, it is quite unitentional and the author would like to add his apologies to his indebtedness.

Many individuals have also contributed to make the book comprehensive and useful to the readers. The author gratefully acknowledges his debt to each of them. The author would also like to acknowledge the support of the entire team of professionals at Stup Consultants Ltd. for encouraging him to convert the idea of writing a book to its fruitful end in the shape of this book. In addition, special thanks are extended to the author's long-time friend and colleague Amitabha Ghoshal of Stup Consultants Ltd. for his close involvement during the preparation of the manuscript as also for going through the first draft and providing invaluable positive suggestions. Thanks are due to T P Nandanan for the long hours he spent ungrudgingly, keying in the entire copy of this book in the computer. D K Mitra who painstakingly typed the first draft also deserves special acknowledgement.

And last, but by no means least, sincere thanks are due to the author's wife, Manjula and son, Indranil for their encouragement and support to write a book on this subject.

15 August 1999   UTPAL K GHOSH

Contents

1

Rationale and Relevance

1.1 INTRODUCTION

Growth of a country largely depends on the development of infrastructural facilities. Transportation facility is one such vital necessity. Roadways, railways, water transport and air transport are the most common modes of transportation. Of these, roadways and railways are the principal forms of transportation for many countries. It is, therefore, natural that new roadways and new railway lines and consequently new bridges are being built as a part of infrastructural development work. The thrust is thus on new development and new construction. As a result maintenance and repair of existing bridges have sadly taken a back seat with consequent insufficient allotment of funds. Apart from this factor, a general misconception that bridges do not require much attention during the first few decades after construction has also contributed to the deterioration of many bridges earlier than expected. However, since these existing bridges are vital assets which contribute to the development and growth of a country the importance of proper maintenance of these bridges should not be allowed to be overlooked. Also, proper maintenance and timely rehabilitation work may well save substantial capital expenditure, which would otherwise be required for replacement of these bridges if left unattended. Rehabilitation of the existing bridges assumes particular significance in this context.

1.2 EVOLUTION OF STEEL BRIDGES

Evolution of steel bridges presents a fascinating study spreading well over two centuries. It will be worthwhile to study this evolution briefly, in order to understand the present scenario better.

Coalbrookdale bridge across the River Severn in the county of Shropshire in England was the world's first iron bridge. Designed by Thomas Pritchard and built by ironmasters Darby and Wilkinson in 1777–79, this 100 ft (30.5 m) span bridge with cast iron arch ribs and using about

400 tons of iron work is still in use after more than 200 years. This was followed by the 130 ft (39.6 m) span cast iron Buildwas bridge over the Severn in Shropshire built by Thomas Telford in 1796. Subsequently, many cast iron bridges were built in Britain and other parts of the world during the next five/six decades. Amongst these the Vauxhall and Southwark bridges over the Thames in London and Pont du Louvre over the Seine in Paris deserve mention.

During the first quarter of the 19th century, good quality wrought iron was being developed. This material which is ductile and malleable soon established itself as a good material for bridge construction. This was recognized by Thomas Telford and in 1826 world's first suspension bridge of 580 ft (176.8 m) main span over the Menai Straits in England was built with wrought iron eyebar chains. This was also the world's first major bridge over sea water. Success of Menai Straits bridge led the way for many more suspension bridges. Amongst these the Hungerford foot bridge over the Thames in London deserves special mention. This bridge was built with wrought iron by Isambard Kingdom Brunel in 1841–1845 with a central span of 676 ft (206 m) and two side spans each 345 ft (105.2 m) long. The superstructure of this bridge was subsequently removed to Bristol and re-used to build the famous Clifton Suspension bridge across the Avon gorge during 1862–1864. This beautiful road bridge is still in use, a century and a half after the original Hungerford foot bridge was built. Wrought iron was also used by Brunel in the bridges across the river Wye at Chepstow (1852) and the Royal Albert Bridge at Saltish over the river Tamar (1859) to carry railways. Use of composite truss bridges with cast iron compression members and wrought iron tension members became popular in the United States during the middle of the 19th century.

In the second half of the 19th century, the technology of steel making was developed by Bessemer and Siemens. Although these early steel materials had properties which were much superior compared to cast or wrought iron, they suffered from one major disadvantage, namely, the quality of the different batches of steel could not be controlled or guaranteed. Research and development continued in different parts of the world, which paved the way for improved technology of steel production and quality control.

1.3 PRESENT SCENARIO

Improvement in steel production technology coupled with advancements of theoretical analysis of structures, technical skill and mechanical appliances heralded the construction of steel bridges towards the closing decades of the 19th century. Thus the mighty Forth railway bridge in Scotland was constructed in 1890 followed by many major bridges in Europe and America. Many smaller and simpler steel bridges were also built during this period. The trend continued in the early decades of the 20th century, which witnessed a marked upsurge in the construction of steel bridges throughout the world including Asia and Africa. Thus many of the existing steel bridges around the world have now become 75–100 years old. These may be showing signs of distress. Also some of these may even be functionally obsolete. Replacing these may appear to be the easiest solution. However, replacement of so many bridges now would mean immediate capital deployment of a gigantic magnitude. Thus there is a clear need for increased attention towards maintenance of the existing bridges. This process would include strategies for inspection, capacity assessment and repair/rehabilitation of the existing stock of bridges. In this context the question of prioritising the repair/rehabilitation process becomes relevant, so that available funds can be utilised in a planned manner.

1.4 BRIDGE MANAGEMENT SYSTEM (BMS)

Bridge Management System addresses the problem of prioritisation of the rapair/rehabilitation process of the large stock of bridges and also helps in formulating strategies for replacement of the existing bridges. The basic requirement of an effective BMS is a strong database consisting of an inventory of all the bridges under the concerned department. This database includes all relevant information about each bridge e.g., location, construction year, environment data, span/width, type of structure, design load, details of sections used in each member, report of past inspections, modifications/repair work carried out or any other special point, etc. In effect the database provides the history of the maintenance and modifications carried out to date. BMS thus calls for regular inspection of each bridge to ascertain the present physical condition of different elements followed by assessment of their load carrying capacities and preparation of strengthening schemes as necessary. This is an ongoing process and gives a clear idea to the authorities about the condition of the entire stock of bridges at any point of time. This information is vital for formulating maintenance/replacement strategy and allotting requisite funds.

1.5 LIFE CYCLE COSTING

Life cycle costing (LCC) is a relatively new concept, which can be suitably incorporated in the Bridge Management System for formulating strategies for maintenance, rehabilitation and replacement of existing bridge structures. This concept considers not only the initial cost of a project, but also the maintenance and other costs to be incurred during the subsequent life of the structure. The basis of LCC is to bring all these future costs to a single value for comparison purpose, often by applying the traditional present value concept on a base date. Thus, if LCC is applied, a solution with higher initial cost may be justified because of reduced future running costs, as compared to a solution with lower initial cost but with high running cost over the total life of the structure. The method can be used for comparison of different solutions using different systems, different materials or different specifications.

Life cycle costing involves four parameters viz., initial costs, future costs, predicted life of the structure and discount rate. While initial costs can be reasonably estimated, there are some misgivings about the accuracy in forecasting the other parameters. Although these perceived difficulties are real to varying degrees, there is no doubt that the method can be utilised as a very useful qualitative tool for decision making. It addresses the maintenance costs in proper perspective and offers an informed and rational way for evaluation of alternative solutions. It also brings to focus that the future maintenance costs are required to be given due importance in the design process itself and that the designers should be aware of the future consequences of their present actions. However, to make it more effective, the designer needs reliable inputs such as databases of maintenance costs and service lives of different solutions as well as an appropriate discount rate.

1.6 STRATEGY

In 1921, J.A.L. Waddell, in his *Economics of Bridgework* stated: The life of a metal bridge that is scientifically designed, honestly and carefully built and not seriously overloaded, if properly maintained, is indefinitely long. This statement is true but the criteria required for attaining an 'indefinitely long life' are hardly fully achieved or achievable in real life conditions. Considering these aspects, many authorities consider 100–120 years to be a reasonable useful life of a steel bridge. However, with modern methods of inspection and timely corrective rehabilitation work, life of steel bridges can be extended even further.

There is thus a clear need of a broad based approach for technical evaluation of the existing stock of steel bridges and wherever possible for developing rehabilitation schemes to upgrade or modify these bridges to suit present day requirement. If a technical solution to rehabilitate and extend the life of a bridge can be worked out, the cost of such repair work, in most cases, would only be a fraction of the cost of building a new one. The money thus saved may be gainfully utilised for other pressing needs. The message, therefore, is very clear. Focus should be on extension of life by rehabilitation rather than on replacement. Therein lies the rationale and relevance of rehabilitation of a bridge.

BIBLIOGRAPHY

1. Berridge, PSA: Factors governing the choice between repairing, strengthening and reconstructing railway girder bridges, *Proceedings of the Institution of Civil Engineers*, London, Paper No. 6702, 1963.
2. Cottrell, AE: *The history of Clifton Suspension Bridge*, Clarks Printing Services Ltd., Bristol (first published in 1928).
3. Chatterjee, Sukhen: *The design of modern steel bridges*, BSP Professional Books, Oxford, 1991.
4. Jones, AEK and Cussens, AR: 'Whole life costing, *The Structural Engineer*, 1 April 1997.
5. Brown, CW and Owens, GW: Whole life costing: ways forward for steel bridges, *Bridge Management 2: Inspection, Management, Assessment and Repair*, Edited by JE Hardinge, GAR Parke and MJ Ryall, Thomas Telford, London, 1993.

2

Inadequacies

2.1 TYPICAL INADEQUACIES

The inadequacies which lead to the necessity of rehabilitation of a bridge may be broadly divided into the following categories:

(a) Some individual members of a bridge may have deteriorated due to natural or man-made causes and may need rehabilitation to bring the bridge back to its original condition. Examples of such problems are corrosion, vehicular collision or accident, effects of war, etc.

(b) Some members which may not have deteriorated, but may need strengthening or modification due to introduction of new loading or design criteria.

(c) There may be yet another category in which a bridge may be structurally safe, but has become geometrically inadequate, because it cannot satisfy the functional requirements of modern traffic. Examples are: wider carriageway requirement, change over to broad gauge railway track or electrified traction system, etc. It may be possible to rehabilitate a bridge with this type of deficiency by incorporating certain modifications only, instead of replacing the entire structure.

It is quite common to encounter more than one of these inadequacies in the same bridge.

2.2 NATURE OF DETERIORATION

Deterioration in steel bridges can be classified according to two broad causative factors, viz. natural deterioration and deterioration due to man-made situations. Examples of natural deterioration are those caused by atmospheric corrosion, earthquakes, floods, etc. Deterioration caused by pollution, stress corrosion, fatigue, accidents, etc. come under man-made situations. Out of these, atmospheric corrosion, stress corrosion and fatigue are the three primary causes of distress. Furthermore, although these are independent mechanisms, they tend to overlap each other, in the sense that any one of these may induce one or both of the other two mechanisms to contribute to the distress. Thus in many cases all the three mechanisms appear to be concurrently responsible for the distress.

Figure 2.1 shows the various causes of deterioration normally encountered. In most of these situations, the extent of damage largely depends on the type of bridge, details adopted, quality of material, workmanship, environment and above all the level of routine maintenance work.

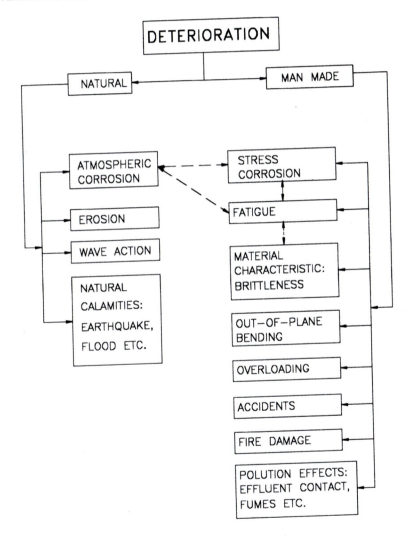

Fig. 2.1: Causes of deterioration

2.2.1 Atmospheric Corrosion

Atmospheric corrosion is an electrochemical process of flow of electricity and chemical changes in steel occurring in stages. The process first takes place locally by formation of anodes (steel substrata), and cathodes (surface mill scales) similar to battery cells. In the next stage, when the surface is moist, the cells start functioning in the presence of an electrolyte media formed out of the moisture film and acidic substances present in the atmosphere. In combination with oxygen from air, hydrated ferric oxide or red rust is formed:

$$4 \text{ Fe} \quad + \quad 3O_2 \quad + \quad 2H_2O \quad = \quad 2Fe_2O_3H_2O$$
$$\text{(iron/steel)} \quad + \quad \text{(oxygen)} \quad + \quad \text{(water)} \quad = \quad \text{(rust)}$$

It is to be particularly noted that corrosion occurs only in the presence of water as well as oxygen. In the absence of either, corrosion cannot occur.

After a period of time, accumulation of the corrosion products (red rust) in the first area causes the corrosion process to be somewhat stifled. This results in the formation of new anodic zone thereby spreading the corrosion area. In cases where the original anodic area is not stifled, the corrosion process continues deep into the steel resulting in pitting or 'pock marks'.

Reduction of area of a member due to atmospheric corrosion increases the stress in the member and indirectly makes the member vulnerable to stress corrosion and fatigue damage.

2.2.2 Stress Corrosion

In corrosive environment, members with high tensile stress are likely to suffer fracture due to stress corrosion. As the cross-sectional area of an already highly stressed member is reduced due to corrosion, the resultant increase in stress may initiate crack. This type of distress is found mostly in specific locations where high concentration of stress is developed, such as eye bars of pins in suspension and cable stayed bridges.

2.2.3. Fatigue Failure

When a steel member is subjected to fluctuation of stresses, as in a railway bridge carrying heavy moving loads, the ultimate strength of the steel is reduced considerably as compared to static load applied gradually. Thus a member may be able to withstand a single application of the design load, but may fail if the same load is repeated for a large number of times. This phenomenon of progressive localized permanent structural change due to fluctuating stresses, that may initiate cracks in the member is termed fatigue failure. This reduction in strength is dependent on two factors viz., number of load repetitions (cycles) and stress variations due to these loads. Fatigue failure occurs at the tension zone of members and may be initiated by stress concentrations at notches, sudden change in cross-sections and sharp corners. Generally fatigue related distresses have been found to be more prevalent in early steel materials with high carbon content. Also, atmospheric corrosion, in many cases, reduces the cross sections of members, resulting in the increase of stress levels, which in a particular cycle of loading may initiate fatigue cracking leading to fracture. These diverging influences on the fatigue behaviour of a structure coupled with lack of sufficiently dependable historical records of number of cycles and stress range already imposed on existing old bridges, prevent accurate assessment of the fatigue strength and fatigue life of these structures. Perhaps one of the most well known examples of stress corrosion/fatigue related failures is the fracture of the eye bar, which caused the sudden collapse of the Silver Bridge over the Ohio river in Point Pleasant, West Virginia, USA in 1967 which took a heavy toll of some 46 lives.

Studies reveal that in welded joints the fatigue strength of steel tends to be reduced due to pronounced changes in the structure (hard grain formation) and properties (lowering of ductility) of the steel. As a result, welded bridges are more prone to fatigue cracks than riveted ones. Consequently, there have been many instances of fatigue cracks in welded steel bridges.

In general, fatigue cracks are more critical in welded bridges than in riveted ones. This is primarily because a crack developed in one component of a riveted connection normally stops at the rivet hole and does not travel beyond to the next component. In case of welded structure, however, crack developed at the weld tends to progress and may affect both the connecting components. Consequently the entire member may be damaged.

2.2.4 Material Characteristic: Brittleness

Brittleness is characterised by a low stress fracture of the material, usually occurring suddenly with little or no prior plastic deformation. Thus, brittle fracture may cause a catastrophic failure.

The key factors causing brittleness in steel are:

- *Metallurgical feature*: Depending on their chemical compositions, heat or mechanical treatment, some steel may be more brittle than others.
- *Temperature:* Structural steel becomes more brittle as the temperature falls. Therefore, fracture may occur even at low stresses, when the ambient temperature drops below freezing point. Geographical location of a steel bridge thus assumes particular significance from this point of view.
- *Service conditions*: Studies suggest that brittleness may be increased due to certain distribution pattern of force fields, such as at locations of stress concentrations due to abrupt change of sections, notches, etc.

Thicker plates normally carry higher chance of brittle fracture due to their complex stress pattern. Other conditions which may induce brittleness include fillet welds laid across tension flanges and intermittent welding causing possible reduction of ductility of the steel in the welded region. Therefore understanding brittle fracture behaviour of steel is particularly important when dealing with welded structures.

2.2.5 Out-of-plane Bending

Distress due to out-of-plane bending may occur where members are subjected to rotation about its longitudinal axis. A typical example of such distress is the rotation of axis of a stringer of a railway bridge at its end connection to the cross girder. During movement of train, the deflection of the cross girder causes repeated rotational movement of the stringer axis. This leads to fatigue related failures. Similar distress may also occur at the connections of cross girders to main girders. However, in this case stiffness of the individual members are normally sufficient to prevent such rotation. All connections of floor beams are thus vulnerable to such distress.

2.2.6 Overloading

With progressive increase in the loading patterns of locomotives, vehicles, etc. the load effects on some existing members or joints of a girder may be found to be more than the capacity for which they were originally designed. Also, in the road bridges, laying of additional wearing course during periodic maintenance work increases the dead load. The cumulative effect of this over a period of time can be quite considerable and is very often overlooked at the design stage.

2.2.7 Accidents

Damages to bridge components due to accidents are quite common. Steel bridges spanning roadways underneath and having inadequate headroom very often suffer impact damages from passing vehicles in the bottom flanges of plate girders or lower chords of truss bridges. Similarly top chords, web members, sway bracings, etc. of through or semi-through type truss bridges frequently get damaged by passing vehicles or due to derailment of railway locomotives or bogies. There are instances of vessels or projections therefrom accidentally damaging overhead bridge structures, while passing along a river underneath. There are also many examples of bridges being damaged due to explosions from war action, sabotage by terrorist activities, etc.

2.2.8 Natural Calamities

Besides these man-made problems, bridges may also be damaged by natural calamities such as floods, earthquakes, etc..

2.2.9 Fire Damage

Generally bridges are not located over highly inflammable buildings or facilities. However, fires may be caused by a rail or truck mounted tanker getting ignited accidentally in the vicinity of a bridge. Accidentally fire may also occur in a temporary or unauthorised store beneath a bridge. All these may cause damage to the structure. The degree of damage depends on two factors viz., the duration and the maximum temperature to which the steel was exposed. Normally steel begins to lose strength at about 200°C and will suffer plastic deformation at temperatures above 620–650°C range. Therefore, whenever any portion of a bridge has been subjected to a major fire, the remaining strength of the affected steel should be duly considered while assessing the capacity of a fire damaged bridge.

2.2.10 Pollution Effect

In highly industrialised areas, chemical corrosion may occur due to presence of chemicals such as chlorides, oxides of sulphur, etc. in the atmosphere, which induces acid films on steel surface causing eventual corrosion. One other example of pollution effect is the corrosion caused by the wastewater discharged from passing trains on the deck steelwork.

BIBLIOGRAPHY

1. Iffland, JSB and Birnstiel, C: 'Causes of bridge deterioration', *Bridge Management 2: Inspection, Management, Assessment and Repair*, Edited by JE Hardinge, GAR Parke and MJ Ryall, Thomas Telford, London, 1993.
2. Owens, GW, Knowles, PR and Dowling, PJ (Ed.): *Steel Designers' Manual* (fifth edition), Blackwell Scientific Publications Ltd., Oxford, UK, 1994.
3. Hammond, Rolt: *Engineering Structural Failures*, Odhams Press Ltd., London, 1956.

3

Rehabilitation Process

3.1 INTRODUCTION

Rehabilitation work for a bridge involves many activities. It starts with the study of the project history and environment of the bridge and after covering many stages culminates in construction implementation at site. A model flow chart of a typical rehabilitation process is shown in Fig. 3.1.

The activities in a rehabilitation process may be broadly grouped under the following heads:

- study of project history and environment
- inspection of existing structure
- capacity assessment
- rehabilitation design
- drawings and specifications
- implementation at site

3.2 STUDY OF HISTORY OF THE BRIDGE AND ITS ENVIRONMENT

The first work for rehabilitation of a steel bridge is to study the history of the bridge and its environments from the available records and drawings. If adequate records or drawings are not readily available, interviewing old employees of related organisations or persons residing in the vicinity of the site of the bridge may provide some valuable information. The date of construction and history of subsequent repairwork or replacement of major members should be ascertained at this stage.

3.2.1 Age of a Bridge

Age provides some idea about the type of steel or iron used in the construction and may influence whether a particular type of repair-work is

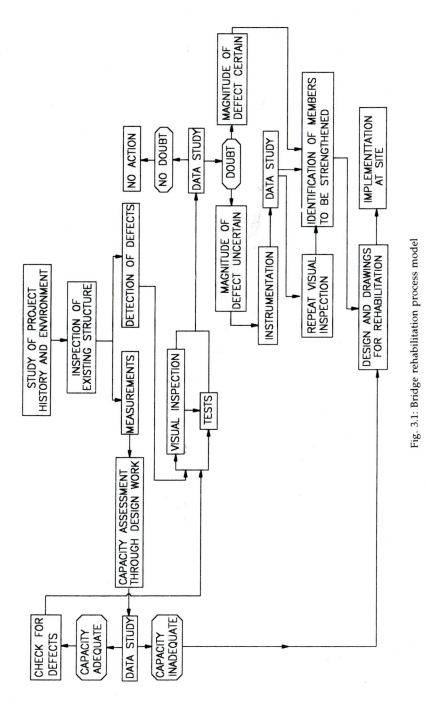

Fig. 3.1: Bridge rehabilitation process model

feasible or not. For example steel with high carbon or silicon content is not easy to weld and has more chance of cracking due to welding than low carbon steel. Similarly, wrought iron is prone to lamellar separation, particularly when heated. Likewise cast iron cannot be welded. In such materials, welding should be avoided for repairwork, although this may appear to provide an easy solution. However, in cases where records or drawings are not sufficient to establish the age or the materials used, modern methods of material testing should provide some idea about the properties of the materials used.

Secondly, age also gives an idea about the loadings and stresses considered for the original design, on the basis of codes of practices prevalent at the time of construction. In many cases, the sections provided may have been more than adequate. Thus modern analysis by computer and design by limit state theory may reveal that despite physical deterioration of some members, these bridges are capable of taking even higher loads than those for which these were originally designed.

Age may also provide a basis for the assumption of the number of fatigue cycles the bridge has already undergone. This information is very useful for assessing the fatigue life of the structure. A point of caution should be mentioned here. The subject of fatigue effect on steel is still under research and an accurate prediction of useful remaining life of a bridge may not be possible. Thus most predictions of remaining life will still remain in the region of 'approximation' only.

3.2.2 Environment of Location

Review of the environment is also relevant for the development of rehabilitation and strengthening schemes. This review starts at the inspection stage and continues till the rehabilitation work at site is completed. This should cover not only the effect of the environment on the existing bridge, but also the effect of the rehabilitation work on the environment.

Presence of water spray or moisture in the vicinity, such as waterfall or marshy wet lands are very often responsible for corrosion of bridge elements. Similarly if the existing bridge has a very low clearance from the highest flood level, the chances of corrosion of the members located at the lower levels become very high. Rehabilitation schemes should take into consideration these environmental hazards and recommend appropriate protective measures. Some hazards are man-made, such as effects of corrosive fumes or chemical effluents from nearby industrial units. These should also be considered in order to minimise chances of recurrence of distress.

History of major fire near the bridge sometimes becomes relevant while assessing the capacity of a bridge. Investigation in this regard is necessary in order to find out whether the intensity of the fire has affected the metallurgy of the steel and rendered it unfit for structural use.

Dumping of debris, release of chemicals, spills of waste materials etc., are the common hazards associated with any construction work. And bridge rehabilitation work is no different. Care should therefore be taken not to disturb the environment while carrying out rehabilitation work. Of particular concern is oil, paint and other harmful chemicals, which find their way to the river water. These chemicals are harmful to fish and other wild life as also to vegetation and may affect indirectly the health of humans as well, if these are used for human consumption. Hazards of human wastes from the rehabilitation personnel are no less important. These aspects should be considered at the planning stage and incorporated in the bridge rehabilitation document itself.

3.3 INSPECTION

3.3.1 General

Understanding the present condition of the various members of a bridge is a prerequisite for any rehabilitation work. Inspection is the first step towards this understanding and in effect the beginning of the rehabilitation process.

Deterioration of any structure is an ongoing process. Ideally therefore inspection and rehabilitation should also be an ongoing process. Thus, every bridge should be subjected to preventive inspections at regular intervals, requiring close examination of all parts of the structure, with a view to identifying and quantifying any deterioration caused in the structure. These findings should be collated and sytematically retained as a database. This database can be used for formulating strategies, such as decision on sequences about rehabilitation of different bridges. Once a decision for rehabilitation of a bridge has been made, a Special Inspection should be undertaken for the specific purpose of developing rehabilitation schemes. This systematic approach would not only assure public safety but also provide data for preventive maintenance and rehabilitation.

Special Inspection may be broadly divided into two activities:

- measurement of existing structure including sections and thickness of all main elements
- detection of defects or deterioration of different members

The former enables capacity assessment of the bridge through analysis and design, while the latter focusses on the defects for the purpose of repair and rehabilitation of individual members.

Out-of-turn Special Inspection may also be required in cases of damage of a bridge due to accident or natural calamity such as flood or earthquake, or after an exceptional load has passed the bridge.

It is not intended to go into details of routine inspection process in this chapter. Guidelines are already available for this purpose. Instead, some basic aspects of inspection which would be useful in the development of rehabilitation scheme will be discussed. The recommendations made and information furnished in this chapter should be useful for both the inspection process and subsequent development of the rehabilitation scheme.

3.3.2 Inspection Personnel

Rehabilitation scheme of a bridge is the outcome of a concerted effort of inspection at site and analytical study at the design office. Ideally, therefore, the structural engineer who is entrusted to develop the rehabilitation scheme should also participate as a member of the inspection team. This would naturally enable him to see for himself the condition of the bridge and understand the deficiency and formulate the rehabilitation strategy much better than going through the report pages prepared by someone else. In actual practice, however, the inspector and the structural engineer may not be the same person. In such cases the inspection report assumes much higher importance, since it is through this report that the structural engineer has to see the condition of the bridge and develop his strategy for rehabilitation schemes.

The problems met with during inspection are rather complex and varied and in many cases difficult to foresee. Thus the inspector's judgement for proper evaluation of the findings at site assumes special importance. Consequently he should be a person closely connected with bridge work and have reasonable knowledge on the behaviour of the structure under actual loading conditions. Additionally, he should be closely acquainted with the construction features of the bridge and capable of putting his findings in clear language and by simple sketches in the report. The inspector should be conscientious and painstaking, and should have developed an insight for observing deterioration of materials due to corrosion, weathering, fatigue, etc. He should be able to identify the areas where problems occur commonly and to initiate preventive maintenance programmes. He should also be expected to recognise the seriousness of any structural deficiency and take appropriate action to keep the structure in a safe condition, if necessary by imposing restrictions in loading or speed or any other measure deemed necessary.

It is seldom possible to select one individual having all the requisite qualifications and experiences of an ideal inspector. It is, therefore, often preferable to form an inspection team of engineers and technicians, with experience or knowledge of diverse areas such as structural design, construction, maintenance, emergency repair, etc. Furthermore, bridge inspection very often involves climbing at heights and not all persons are good at it. This aspect should also be considered while forming the team for inspection. Assistance of specialist engineers/agencies should be sought in cases of large or badly damaged bridges, movable bridges involving mechanical and electrical equipments, or special structures such as suspension or cable stayed bridges.

3.3.3 Areas to be Inspected

Although each and every element of a bridge structure needs to be inspected, there are some areas where serious defects are likely to occur and therefore need particular attention during inspection. Some of these areas are indicated below:

1. Members with high design stresses
2. Members where water may collect due to inadequate drainage system, such as main joints and U shaped bottom chords.
3. Areas which are subject to alternate wetting and drying, such as areas under timber sleepers in railways bridges
4. Areas which are not easily accessible to painting, such as troughs in ballasted deck system, areas adjoining interfaces between RCC deck slabs and supporting girders, areas behind the bends of joggled stiffeners, etc.
5. Areas of steel girders where waste water is discharged from passing trains in railway bridges or areas directly below drainage spouts or weep holes.
6. Areas which are prone to corrosion due to emission of steam blasts or smokes of locomotives.
7. Areas (particularly undersides of bottom chords) where dews collect or where water from the river touches during floods.
8. Surfaces of the windward side of a bridge situated near Sea or over creeks which are prone to more corrosion than other areas.
9. Joints of deck system in general, particularly:
   - throats of cleat angles or seating angles
   - notches in the flange/web at the end of stringers and cross girders (for fatigue crack)
   - loose rivets/bolts
   - corroded cleats and rivets/bolts
10. Top and bottom lateral bracings which are liable to be bent or buckled. Also the gussets at the ends and at the middle of bracings.
11. Web members of through trusses, particularly portals and sway bracings, as also the lower members of girders of road over-bridges for damage due to collision with vehicular traffic.

12. Camber of truss
13. Compression flanges, web plates and stiffeners of plate girders as also compression members of trusses for signs of deformation or buckling
14. Unusual vibration or excessive deflection under passage of heavy load should be noted and the cause investigated
15. Creep or logitudinal movement of girders
16. In welded girders fatigue prone areas should be carefully inspected. Special care should be taken to inspect areas where there is abrupt change in the size of metal or in configuration which may produce an area of concentrated stress, or in areas were vibration of movement could produce stress concentration. Likely locations of fatigue cracks are: end of cover plates in flanges with multiple plates, connections of flange plates of different thicknesses, welded splices, ends of stiffeners, intersections of longitudinal and vertical welds in webs, etc. Also defects in welds such as excessive porosity, lack of fusion, notch or under cutting may cause increased stress and initiate cracks. Fillet welds across the direction of tensile stress are prone to fatigue cracking. Stress raising details such as notches and corners should draw special attention.
17. In bearings, the common examples of distress are bent or loose holding down bolts, frozen rollers, cracks in base plates. The bearings should be checked for tilt and also whether all the bearings are at the same level. Problems generally encountered in bridge bearings have been discussed in detail in Chapter 4.
18. While carrying out inspection, special attention should be given to 'fracture critical' members of a structure. The subject of redundancy and fracture critical member has been discussed in a later section of this chapter (Section 3.5.4).

3.3.4 Inspection Tools

Some of the most useful tools which should accompany any inspection team are a 2 m pocket tape, a 30 m steel tape, chipping hammer, paint scraper, wire brush, plumb bob, vernier or jaw type callipers, small level, steel straight edge, rivet testing hammer. Other items such as flashlight, mirror, magnifying glass, a set of crayons, nylon chord and gloves may also be useful during inspection. A clip board is also essential for taking notes.

3.3.5 Inspection Equipment

Inspection of members whether situated above or below the deck will need some equipment for access. For structural elements at or above deck level simple equipments such as ladders, portable platforms, planks, etc., should suffice. However, where access is not so easy, special equipments have to be considered. Some equipments which are normally used for inspection purpose are briefly described in the following paragraphs:

(a) Temporary Scaffolding

This is the most common method of bridge inspection, particularly in cases where this can be constructed from ground level or where this does not interfere with ongoing traffic. Scaffoldings may also be suspended from the bridge structure itself, in which case particular care should be taken not to infringe the minimum clearance dimensions. Temporary scaffolding is, however, time consuming to erect and dismantle and consequently relatively expensive. Inspite of this disadvantage, however, temporary scaffolding is the most popular inspection equipment, particularly for small bridges situated in remote areas.

(b) Travelling Gantries

Some long span bridges are equipped with custom-made permanent travelling gantries suspended from the structure and capable of travelling along the length of the span.

(c) Lifting Platforms

These are platforms mounted on vehicles and can be lifted or lowered vertically by hydraulic or manual method. The entire unit can be moved horizontally and used for inspection of members at a height of 10–11 m. These are fairly cheap, but need reasonably level ground below a structure.

(d) Boats with Platforms

These are quite easy and safe equipments, particularly for inspection of the underside of the bridge members.

(e) Bucket Snoopers

Hydraulically operated truck or rail-wagon bucket snoopers allow access from bridge deck to the sides or underside of the bridge. These bucket snoopers have an added advantage of threading through truss members.

(f) Truck Mounted Telescopic Hoists

Where access from roadway below is available, truck mounted hydraulically operated telescopic hoists fitted with bucket may be conveniently used.

3.3.6 Testing Methods

There are many modern non-destructive testing (NDT) methods which may be utilised by the inspection team for ascertaining actual thickness of steel members in hard-to determine areas, such as girder web or for investigation of various faults such as cracks, laminations, etc. Some to these methods are:

(a) Thickness Measurement

Thickness of members subject to corrosion can be measured by direct measurement with the help of callipers where access is available. In case access is not available, this can be done by ultrasonic thickness gauge. This is a very handy equipment which enables the thickness to be measured from any one surface to an accuracy of approximately 0.1 mm. The gauge usually gives digital reading and is very useful, particularly for ascertaining the corrosion loss in thickness of a member for assessment of remaining strength. Special care should be taken to clean any rust or loose paint from the surface before taking measurement. In case of pitted surface, it may be necessary to spot-grind the area, to ensure good contact between the probe and the metal surface.

(b) Crack Testing

Cracks in steel can be discovered by using several non-destructive testing methods. Some of the more common methods are briefly described below:

Dye penetrant testing: This form of testing is widely used at site and can be readily used for detection of surface cracks. In this method the surface area is first cleaned of any rust or paint before the dye penetrant is applied by spraying. The dye seeps into any cracks or other defects open to the surface. After the penetration time (about 20 min), the excess penetrant on the surface is cleaned off using a solvent. A developer (like a chalk powder) of contrasting colour with high absorbant property is then applied. If any cracks or surface defects are present, the dye bleeds out of the defects and appears as stain on chalk surface. The surface is then examined by using a magnifying glass. This method of testing is relatively convenient for use at bridge sites, but proper care should be taken for preparation of the surface.

Magnetic particle testing: This method makes use of the magnetic properties of steel by setting up a magnetic field within the piece to be tested and dusting the test area with fine iron powder or spraying the test area with liquid detection medium carrying magnetic iron powder. If there is a crack, the magnetic lines are disturbed and a small north-south pole area at the surface crack zone is created and outline of the crack becomes clearly visible. This method is suitable for detection of surface or subsurface cracks. There are a few techniques of producing the magnetic field in the test area. In the first type, the poles of a strong permanent magnet or an electromagnet are placed on either side of the area to be tested. In the second type electric current is passed through the steel sample itself. In the third method, magnetic field is produced by induction by winding a coil around the test area and passing a current through the coil.

Ultrasonic testing: This method is widely used for detection of surface and subsurface flaws in steel, such as cracks in welds, laminations in plates, etc. The testing involves the introduction of a high frequency sound beam into the area to be tested by means of an ultrasonic transducer. The sound beam travels through the steel until a crack is met, when it reflects back to the transducer. This produces a voltage impulse which appears on the cathode ray tube (CRT). This test is very sensitive and has the advantage that it requires access from only one side of the material (unlike radiographic tests). The test can also be conveniently carried out at bridge site, as portable testing machines are available. However, the method does not provide any permanent test record. It also requires specialised skill in interpreting the pulse-echo patterns appearing on the screen.

Radiographic examination: This method may be used for detecting both surface and subsurface defects. The system essentially consists of passing X-rays or gamma rays through the member being tested and creating an image on a photosensitive film. If there is a crack in the member or a void in the weld, less radiation is absorbed by the steel and more radiation passes through that area to the film. Thus the defects are shown on the film as dark lines or shaded areas, compared to the surrounding areas of sound material. This method has an advantage of providing a permanent record for every test carried out. However, it requires specialised knowledge in selecting the angles of exposure and also in interpreting the results recorded on the films. It also requires access from both sides of the test area, with the radiation source placed on one side and the film placed on the other side. One other point needs special mention here, viz. the safety aspect, because the system uses radioactive isotopes requiring certain statutory safety precautions.

Hardness testing: It may be necessary to test the hardness of steel material in a particular location at site. For this purpose portable hardness testing equipment may be conveniently used. There are two types of such equipment. The first type is based on ultrasonic properties, whereas the second type utilizes rebound principle. Hardness testing is particularly relevant in case of fire damage, for ascertaining whether there has been any significant change in the property of steel due to fire, by comparing the test results from the areas close to and remote from the fire zone.

Coupon testing: It may be necessary to remove small samples (coupons) by cutting or drilling from the steel structure itself, in order to carry out laboratory tests to establish physical and chemical properties of the steel used in a bridge. However, utmost care should be exercised in selecting the locations so as not to affect or impair the capacity of the member or stability of the bridge. In case it becomes absolutely necessary to take specimens from main members, care must be taken to ensure that removal of samples does not permanently weaken the structure, and if necessary, a bolted repair detail should be introduced. In any case, the Engineer's approval should always be obtained prior to removal of any coupon from the structure.

3.3.7 Field Load Test and Instrumentation

In some special cases the behaviour of a bridge may be examined by subjecting it to actual or simulated design loadings and observing the effects on the critical members by instrumentation. Prior to loading, strain gauges are fixed on critical members. The structure is then loaded by placing trucks or locomotives of known weights at various points of the bridge and the strain is recorded. These tests are normally done with incremental loads. When static load tests are conducted, due allowance for impact effect should be made. Based on the strains at different locations, the stresses in various members are worked out and are compared with the analytical data obtained by computation.

Testing may also be carried out by passing the test vehicles over the bridge at incremental speeds starting from static position. In such tests, apparent vibration during high speeds, as also behaviour of bearings, bowing of compression members or any tendency of observed cracks to open under load may be of importance for understanding the overall behaviour of the structure under normal traffic loads.

In general, field load test can be carried out in the following stages of operation:

- fix strain gauges at selected locations
- apply the loads at different positions to maximise the stress in different members
- compute stresses (and forces) from measured strains on the different members

A typical arrangement of instrumentation for a truss bridge in shown in Fig. 3.2.

Load induced deflections (both vertical and sideways) at specific locations of a bridge structure can be measured by means of suitable survey instruments. The measurements can be taken by using dial gauges for direct visual reading, or electrical systems where remote recording is needed. Laser techniques are sometimes utilised for measurement of deflections. For measuring rotation or change of slope at some strategic points, special equipments (inclinometers) may be conveniently used. There are a number of methods for measuring deflection. It is, however, necessary in all cases to establish firm datum points for measurement of these deflections.

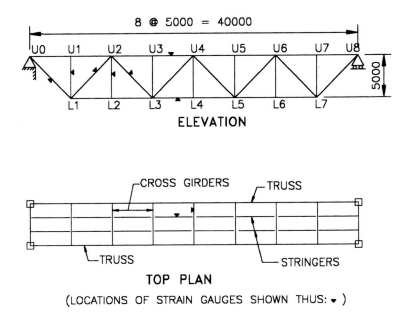

Fig. 3.2: Typical arrangement of instrumentation in a truss bridge

Also, periodic checks should be made to ensure that there is no movement of these datum points during the measurement process.

3.3.8 Safety

At all stages during inspection, top priority must be given to safety. It is advisable to draw up a safety programme well in advance of the proposed inspection activity. The programme should cover:

(a) the safety and welfare of the persons at work and
(b) protection of other people including members of the public against risks to health and safety arising out of the activities at work.

This programme should cover standard bridge inspection safety procedures of the concerned authorities as well as other safety requirements, such as traffic control procedures to conform to local regulations (e.g. warning signs, flashing arrow boards, cones, etc.). Every member of the inspection team should be equipped with safety vests, hard hats or helmets, work boots, etc. Safety belts should be used if climbing is necessary. Busy areas may require night time work, which may warrant special precautions, particularly to the inspection personnel. A first aid kit box must accompany the inspection team.

3.3.9 Photography

A camera is an excellent tool for a bridge inspector. Clear and sharp photographs should always form part of the inspection report. Modern cameras fitted with wide angle/telephoto lenses are of immense value during inspection. Wide angle lens is useful for taking close-up details or overall view of an entire area, where the space for going back far enough for the purpose is limited. On the other hand telephoto lens or cameras with zoom lens can be gainfully employed to take close-up details of the subject from a distance, when the area is not easily accessible. Photographs should always include the location of the details written in chalk on the structure itself. It is also advisable to include in the photograph a clearly marked scale, or an easily recognisable item for easy comprehension of the scale of the detail.

3.4 CAPACITY ASSESSMENT

Capacity of a span relates to the net strength of the individual members available for live loads and impacts, after deducting the requirements for dead load. The dead load should include not only those for the existing bridge but also the estimated load for the additional materials for repairs or strengthening. These estimates should be fairly accurate, as otherwise the assessed remaining capacity for live load may be inaccurate. It is necessary to calculate the capacity of all the critical members of a bridge and compare these with the actual live load effects on these members. This will help to identify the members which are deficient and require strengthening.

Capacity assessment can be done either by traditional working stress method or by the more modern limit state theory. Use of computer and application of limit state theory may even reveal that the capacity of the structure is higher than that calculated manually by traditional methods earlier.

3.5 REHABILITATION DESIGN

3.5.1 General

Having identified the individual members to be strengthened from the results of inspection and capacity assessment, the next activity is rehabilitation design. This activity is carried out broadly in two stages, viz. concept stage and design stage.

3.5.2 Concept Stage

The concept stage is the most important one as, during this stage the problems and the various options for rehabilitation are considered in detail leading to the next stage of detail design. A few relevant points need to be considered prior to the examination of the different options.

(a) It is imperative to have a very clear understanding not only of the problem, but also of the cause of the problem. Unless the cause is known the effective remedy cannot be prescribed.

(b) While developing alternative schemes, the practical aspect of implementation of the solution should be of paramount importance. It will be futile to arrive at an economic solution which cannot be implemented at site. An example of such a situation is to propose bolted connection of a new member, where there is no approach to fix and tighten the bolts. Also a solution with the use of special equipment should be avoided as far as possible, as this is likely to increase the cost and limit the number of bidders.

(c) Rehabilitation work is generally labour intensive and cost of new material input is only a fraction of the total cost. Therefore it is preferable to select a solution which is inherently straightforward to implement and needs minimal or no special equipment or specialist labour force, even at the cost of higher material input. In fact, providing some extra material may be advantageous if considered at a wider perspective, since, this may supplement any deficiency which may not have been forseen at the design stage. Often this is quite useful, as load paths in existing structures with some damaged members are not always easy to predict and therefore the actual forces may exceed the calculated forces. Extra material will stand in good stead in such an eventuality.

(d) One other point needs to be considered during concept stage. The preferred rehabilitation scheme should be such as to use, as far as possible, partially shop fabricated units, which may be supplied to site in knocked-down condition, to be assembled and fixed at site. This enables the repair work to be done with minimal disruption of traffic.

3.5.3 Rehabilitation Methods

During concept stage as many alternative solutions as possible should be developed in writing and/or by preparing sketches. Modern approaches like brain storming sessions and lateral thinking are some times very helpful in evolving the most suitable solutions. The various solutions may be subsequently reduced to two or three viable solutions considering economy, simplicity, available resources, etc.

Although the problems encountered in each bridge are to some extent unique and cannot be generalised, the following paragraphs describe some of the common methods which may be considered for developing conceptual schemes:

(a) Repair of Critical Members
If a particular member is found to be deficient, or damaged, this may be strengthened by adding suitable material on to it. Alternatively the entire member may be replaced by a new member of the required strength. Care should, however, be taken to check that the connection details are sufficiently strong to transmit the required forces. Very often the connection details play important role in such repair work. In case of local bending of a member, it can be strengthened by mechanical method, or by means of flame straightening.

(b) Introduction of New Member to the System
A deficient structure can be rehabilitated or upgraded by introduction of new structural elements to it. An example of this procedure is to add additional stringers between the existing ones to increase the load carrying capacity of an inadequate deck system in a bridge. Care should, however, be taken to consider the deflection of the new members *vis-a-vis* that of the existing members to ensure effective participation of the new beams of sharing the load along with the existing stringer beams.

(c) Reduction of Dead Load
Reduction in the dead load of a bridge increases its live load carrying capacity. This aspect should be explored while preparing the conceptual schemes. One example of this method is to replace ballasted railway deck system by sleeper system, which would considerably reduce the dead load of a railway bridge. Similarly, in a roadway bridge the dead load of a reinforced concrete (RC) deck system can be reduced by replacing the RC deck slab by orthotropic steel deck system.

(d) Modification of Structural System
The capacity of a bridge can be increased by modifying the basic structural system. For example, simply supported spans of longitudinal beams of the deck system may be converted to continuous beams, by suitably modifying the end connections of these beams. Similarly existing non-composite RC slab on steel girders in the deck system may be changed to act compositely by fixing new shear connectors on the steel beams. Adding new supports to a bridge will reduce the span and thereby increase the capacity. One other example of modification of structural system is by introducing counterbalancing forces in the system by prestressing techniques.

Examples of rehabilitation by the above methods will be discussed in greater details later in the book (Chapter 7). In some cases a combination of the above schemes may be employed to achieve the required results.

3.5.4 Design Stage

Once a few viable conceptual schemes are identified, these are subjected to more rigourous analysis and design work. During this stage a few important aspects need to be considered.

(a) Dead Load Stresses

Once a bridge is erected, the members become stressed due to dead load effect. It is desirable to relieve the members of this dead load stress prior to undertaking rehabilitation work. Otherwise the existing members would continue to be stressed to the extent of dead load effect. Consequently the capacity of the new material will remain under-utilised, as this cannot reach permissible stress level without overstressing the existing members. Where it is not practicable to relieve any dead load, the new material should be considered to carry the live load only. In any case, proper analysis should be carried out to avoid overstressing of the existing members.

When a member suffers local buckling due to, say, vehicular collision and is straightened by heating or by mechanical device, residual stresses will be locked into the member. To avoid this, wherever possible, the damaged member should be taken out from the structure, then rectified and finally reerected into the structure. If the damage is beyond repair, the member should be replaced by a new one. However, when a member is removed for repair or replacement, the effect on the existing structure due to this removal should be examined carefully as the sharp increase in dead load stresses in the adjoining members might cause sudden displacement of the joints of the concerned member. To avoid this condition, relief of dead load stresses in the structure should be considered prior to removal of the member. There are a few methods for relieving dead load stresses of an existing bridge girder while carrying out rehabilitation work at site. The most common method is to jack up the girder and provide temporary supports under the entire structure at a number of points. Another method is to erect a temporary girder above the existing bridge and to provide supports from the top at a number of points by means of hanging device fitted with adjustable screw arrangements. This latter arrangement is suitable where temporary supports under the bridge is not possible. Figure 3.3 shows some typical methods for relieving dead load stresses in a bridge structure.

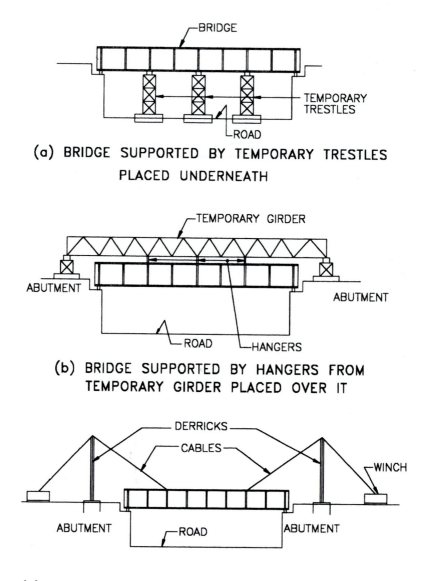

(a) BRIDGE SUPPORTED BY TEMPORARY TRESTLES
PLACED UNDERNEATH

(b) BRIDGE SUPPORTED BY HANGERS FROM
TEMPORARY GIRDER PLACED OVER IT

(c) BRIDGE SUPPORTED BY SUSPENDER CABLES FROM
TEMPORARY DERRICKS PLACED AT ABUTMENTS

Fig. 3.3: Methods of relieving dead load stresses

(b) *Redundancy and Fracture Critical Member*

A redundant bridge structure has, within itself, multiple load carrying mechanisms, so that in case of failure of one mechanism, the load will be carried by another. A non-redundant structure on the other hand does not have multiple load carrying system and consequently failure of a single element may cause collapse of the structure. This element is generally termed 'fracture critical member'.

Redundancy can be broadly divided into three categories, viz. load path redundancy, internal redundancy and structural redundancy.

Load path redundancy is achieved when there are sufficient numbers of load carrying members, so that even if one member fails, the load would automatically redistribute itself amongst the other members and the structure as a whole would still not collapse. Examples of load path non-redundant structures are a two-plate-girder bridge, or a two-cable suspension bridge. A multi-stringer bridge may be considered as a load path redundant structure, whereas a three girder bridge may or not be a load path redundant structure.

Internal redundancy is achieved when members are composed of multiple elements and failure of one element does not cause collapse of the structure. For example a riveted bridge, where individual members consist of sections joined together by rivets, may be considered as internally redundant. Therefore, although a two-truss or two-plate-girder riveted bridge may be a load path non-redundant structure, it is an internally redundant structure. However, a welded two-girder bridge may not be considered either a load path redundant or an internally redundant structure.

Structural redundancy is best exemplified by a bridge structure with continuous spans, where even if a beam fails, it still carries the load, because of continuity.

It is necessary to clearly understand the difference between the three categories of redundancy. The interpretation may vary due to the type of design adopted, such as a two-girder or multiple-girder system, simply supported or continuous structure, riveted or welded construction, etc. Identification of the fracture critical member therefore requires an in-depth study.

(c) *Effects of Fatigue*

Fatigue effect should be examined properly during the development of rehabilitation details, particularly in the fracture-critical elements. Some of these details which need particular attention include transitions at ends of cover plates, welded connections in tension members and stress raisers such as notches, sharp corners, rivet or bolt holes.

While developing a welded design for retrofit, a few recommendations which would help in reducing the adverse effects of fatigue are indicated below:

- Butt welds are preferable to fillet welds
- Double sided fillet welds are preferable to single sided fillet welds
- Fillet welds across the direction of stress should be avoided
- Effects of localised stress concentration factors should be considered
- Intersections of longitudinal and transverse welds should be avoided
- Welding procedures should be carefully chosen and finalised
- Suitable non-destructive testing (NDT) should be adopted

(d) Types of Connections

Defective or loose rivets should be replaced by close tolerance or friction-grip bolts. These new bolts should carry both dead and live loads along with the existing rivets and the joints should be checked as such. Welding on existing rivetted connection should be avoided, as the welds are likely to take the entire load in case there is any slippage in the rivetted connection. Even if the welds are designed to take such load, the weldability of the parent material must be ascertained by laboratory tests.

In any case, fillet welds should not be allowed across the direction of stress in tension areas or areas subjected to stress reversals, to avoid adverse fatigue effects.

In case of welded bridges, additional plates or sections may be welded to the existing members. However, care should be taken to detail the joints properly and supervise the field welds closely. There are instances of damage to bridge caused by indiscriminate and improper welding carried out during rehabilitation process. It would therefore be advisable to develop a scheme with bolted connections instead of welded connections.

Various aspects related to selection of fasteners and analysis of typical connections have been discussed in greater detail in Chapters 5 and 6.

(e) Effects of Eccentricity

It is a common practice to strengthen a member by adding additional plate or section on to it. In such a case details should be developed to ensure that the centre of gravity of the strengthened section coincides with the centre of gravity of the original section. In case it is not possible to achieve this, the effects of eccentricity should be considered in the design.

3.6 DRAWINGS AND SPECIFICATIONS

Since rehabilitation work involves working on existing members and connections, it is imperative that the drawings and specifications prepared by the engineer should be clear and unambiguous. All necessary details and preferably the installation sequence of the proposed rehabilitation scheme should be clearly indicated in the drawings and specifications. In case the contractor is required to produce his own working drawings, the engineer must thoroughly check the drawings and the proposed sequence of installation to avoid costly delays during actual work at site. The working drawings must take into consideration the actual measurements at site, based on the engineer's conceptual drawings. The specifications should also be very exhaustive and should cover as many problems as possible, which may come up at the time of execution.

3.7 IMPLEMENTATION AT SITE

Implementation at site is the final culmination of a rehabilitation scheme. Thus, successful and satisfactory rehabilitation work depends as much on the quality of work at site as on the theoretical work in the office. While many of the activities at site are similar to those for a new construction, there are some aspects which need particular attention during rehabilitation work. These aspects are discussed in the following paragraphs.

Implementation of repair and rehabilitation scheme for a bridge is mostly a time-bound operation. It is therefore very necessary to make a detailed study of the erection scheme and anticipate any possible problem that may arise during erection, so that timely action can be taken beforehand to ensure completion within the target date. The action plan should consider the infrastructure and inputs such as crane facility or lifting device and other equipment required for executing the work. Monitoring of this programme along with close technical supervision at every stage of site work is very important. Also, inspite of the designer's best efforts, there may be problems which have not been envisaged earlier and the supervisory team will often be called upon to solve these on the spot. Therefore, the team at site should be fully equipped to deal with such contingencies.

Workmanship at site will need particular attention. Fabrication and erection of every part of the work should be done most accurately, so that the parts fit properly together on erection. The parts that do not fit should not be bent, forced or hammered into place. Pressure applied for straightening a steel material should be such as not to injure the material. Where heating is used for straightening, the temperature of the steel should not exceed 650°C. Also accelerated cooling should not be done. After the member is straightened, the surface of the metal should be checked for evidence of fracture.

Bending of steel plate by cold process should be done carefully, so as not to injure the material. As a thumb rule, for cold bending, the internal radius of bends should not be less than 2.5 times the metal thickness for plates up to 25 mm thick. As in the case of straightening, the surface of the metal should be carefully inspected to detect any fracture due to bending.

Dismantling of an existing member, if required, should be done without any damage to other members or adjacent steelwork. Flame cutting of existing steels should be done carefully by experienced operators without overheating the adjoining areas. Many of the existing bridges are of riveted construction. Rehabilitation work on these bridges often entails removal of rivets. This work requires utmost care. Rivets may be removed by first shearing the head by using a pneumatic rivet breaker and then driving out the rivet shank with a pneumatic punch. Alternatively, the rivet head may be flame cut about 2 mm above the base metal and the shank may be driven out by means of a pneumatic punch. Rivet shank may also be driven out by drilling in both these cases. Whichever method is employed, care should be taken to ensure that the parent metal is not damaged.

BIBLIOGRAPHY

1. Bakht B *et al.*: *Repair and Strengthening of Old Steel Truss Bridges,* American Society of Civil Engineers, New York, USA, 1979.
2. *Bridge Inspection Guide:* Her Majesty's Stationery Office, London, 1983.
3. Feuer, NJ and Little, RG: 'Bridge Inspection', *Bridge Inspection and Rehabilitation*, Parsons Brinckerhoff, John Wiley & Sons Inc., USA, 1993.
4. *Manual for Maintenance Inspection of Bridges:* American Association of State Highway and Transportation Officials (AASHTO), Washington D.C., USA, 1974.

4

Bearings

4.1 INTRODUCTION

Although small in size compared to other bridge elements, bearings are of significant importance for the proper functioning of any bridge. They transmit the vertical and horizontal forces from the superstructure to the substructure. They are also required to provide movement of the bridge span caused by variations in temperature. It will be useful for the engineer designing a rehabilitation scheme to have an indepth understanding of the functional requirements of bearings before going into the potential problems and their remedial solutions. Also, prior knowledge of the types of bearings available in the market and their functions, advantages, limitations, etc. gives him an added advantage of examining the problems with an open mind without restricting his vision to the existing bearings only. This is particularly significant for the design of replacement bearings when, apart from the usual design considerations, the actual causes of distress for the particular bearings are required to be addressed. It is quite possible that the problems faced by the existing bearings can be largely eliminated if, instead of using the same (existing) type of bearings, more appropriate modern bearings with wider range of functional flexibilities are used for the particular bridge.

4.2 FORCES ON BEARINGS

Bearings are called upon to withstand a great variety of forces. The basic forces are as follows:

4.2.1 Vertical Forces

These are mainly the dead load of the bridge components and live load from the passing vehicles.

4.2.2 Transverse Forces

These consist of wind, earthquake and other horizontal forces acting across the centre line of the bridge and are generally resisted by keys and anchor bolts of the bearings. Bridges on horizontal curve may be subjected to significant outward transverse force due to centrifugal effect from vehicles passing at high speed or applying brakes on the curve.

4.2.3 Longitudinal Forces

These forces include traction and braking forces, thermal forces, forces due to shrinkage of concrete, etc. and act parallel to the centre line of the bridge.

4.2.4 Uplift Forces

These relate to reactions from transverse forces such as wind or earthquake or from centrifugal effect of passing vehicles in locations of sharp horizontal curve.

4.3 MOVEMENTS IN A BRIDGE

Generally, the movement in a bridge is in the longitudinal direction. However, sharp skewed bridges or very wide bridges do experience transverse movements as well.

Change in temperature is by far the most potent cause of longitudinal movement in a bridge. Throughout the life of the structure, bridge components expand when heated and contract when cooled. Bearings are required to allow for these movements. If such movements are stopped or constrained, the bridge structure may be subjected to considerable additional forces, even in case of small spans.

4.4 TYPES OF BEARINGS

The bearings generally found to be in use in the existing bridges may be classified into two broad categories, viz., fixed and expansion bearings. Fixed bearings resist longitudinal movement but allow rotation, induced by the deflection of the span. Expansion bearings accommodate both longitudinal movements as well as rotation. Both the fixed and expansion bearings are designed to transmit vertical as well as transverse forces to the substructure. Before going into the problems commonly encountered in the bearings of existing bridges, it will be useful to briefly discuss the components and functional characteristics of some of the common forms of bearings which are generally used in bridges.

4.4.1 Plate Bearing

This is the simplest and perhaps the oldest form of bridge bearing and essentially consists of two steel plates, one (bearing plate) fixed to the bridge structure and the other (bed plate) fixed to the bed block, sliding at their interface to accommodate horizontal movements. In some plate bearings bronze plates or lead sheets are provided at the interface to reduce the coefficient of friction. Alternatively, grease or oil are also used as lubricant. Plate bearings are used for smaller bridges (12 m to 20 m spans), where rotation at ends due to deflection of the girders may be neglected. In some cases, however, the upper bearing plate is bevelled to allow for rotation of the girders. Figure 4.1 shows a typical arrangement of the latter type of plate bearings.

4.4.2 Rocker and Roller Bearings

For spans beyond 20 m where the deflection of the girder due to live load becomes significant, rocker and roller bearings are commonly used to accommodate both the horizontal and rotational movements at supports. These may be of either cast iron or cast steel for older bridges and mild steel fabricated construction for more recent ones. There can be many varieties of these bearings.

Figure 4.2 shows a typical rocker type fixed bearing, where rotational movement is provided by a solid shaft circular machined pin inserted between two pin plates with machined semi-circular recesses placed above and below the pin.

Figure 4.3 shows a typical roller type expansion bearing, which allows longitudinal as well as rotational movements. While the rotational movement is provided by saddle-knuckle arrangement, the rollers below the knuckle slab provide for the longitudinal movement. Vertical tooth bars on two sides of the rollers prevent lateral movement of the span. For multiple roller system it is necessary that the rollers remain parallel and uniformly spaced during movement, otherwise these are likely to get jammed. Generally, horizontal guide plates connecting all the rollers are provided for maintaining correct alignment. Also, since only the top and bottom surfaces of the rollers are used for movement, the unused (superfluous) side portions of the circular rollers are often eliminated in the more recent bearing forms. This not only saves material but also provide spaces for accommodating multiple rollers in a bearing.

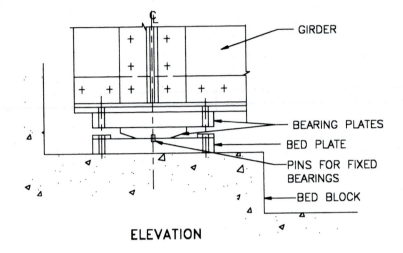

ELEVATION

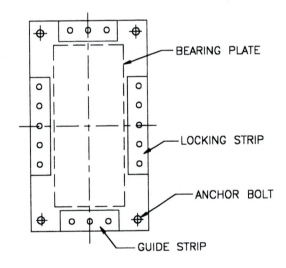

PLAN AT BED PLATE LEVEL

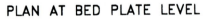

Fig. 4.1: Plate bearing

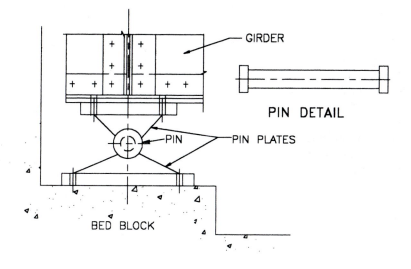

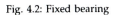

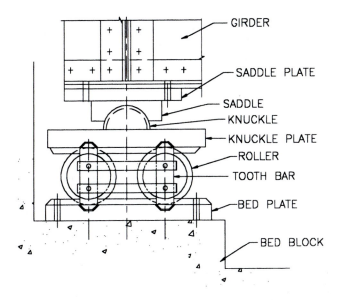

Fig. 4.2: Fixed bearing

Fig. 4.3: Expansion bearing

4.4.3 Elastomeric Bearings

Elastomeric bearings have gained popularity in recent years. These bearings are made of elastomeric materials (natural rubber or synthetic materials like neoprene) generally placed between top and bottom steel plates. It is a common practice to introduce steel sheets between the layers of elastomer to prevent outward bulging of the material due to service load. These steel sheets separating the elastomer are also completely encased within the elastomeric material. Figure 4.4 shows a typical elastomeric bearing. These bearings utilise the characteristics of the elastomeric material, to cater for longitudinal movements as well as rotations at the ends and provide virtually maintenance free service, because of the weather-resistant behaviour of the material and absence of moving parts.

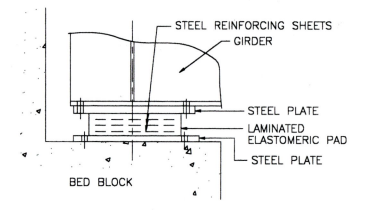

Fig. 4.4: Elastomeric bearing

4.4.4 PTFE Bearings

PTFE is a short term for poly-tetra fluoro ethylene and has an extremely low coefficient of friction with stainless steel surfaces. Use of PTFE in providing sliding surfaces on bearings has been a very recent development. Essentially a thin stainless steel plate is seam welded to the top bearing plate. The stainless steel plate bears on the PTFE sheet placed on the bed plate and provides a very effective sliding surface. Since bonding property of PTFE is very poor, it is customary to locate PTFE by confinement all around to prevent sliding away from the bearing. Figure 4.5 shows a schematic arrangement of a typical PTFE bearing. PTFE has also been used for sliding bearing in combination with elastomeric bearings. Thus while the rotation is accommodated by the elastomeric material, the lateral movement is taken care of by PTFE.

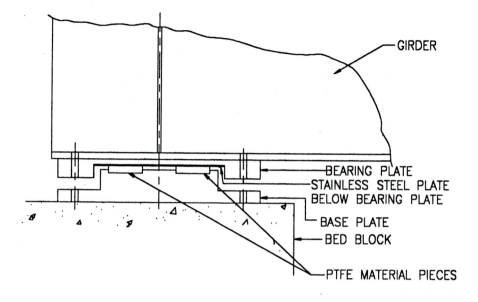

Fig. 4.5: PTFE bearing

4.4.5 Pot Bearing

Pot bearings are also of recent development and are used as an alternative system for heavy type traditional steel bearings accommodating both rotation and lateral movement. Essentially a pot bearing consists of a circular elastomer pad which is confined within a steel cylinder which prevents it from bulging and enables it to carry more load than it would have normally done, if allowed to bulge. The vertical load is transmitted to the elastomeric pad by a circular steel plate acting as a piston inside the steel cylinder. A metallic (commonly brass) sealing ring is provided inside the cylinder to prevent leakage of the elastomer from the clearance between the piston and the cylinder. The elastomeric pad provides the required rotation. To accommodate horizontal movement, PTFE sheet is normally fixed on top of the circular steel plate while a stainless steel plate is seam welded to the underside of the bearing plate. A pot bearing can be conveniently used where rotation and horizontal movements in more than one direction are required to be catered for, such as in a curved girder bridge. Figure 4.6 shows the arrangement of a typical sliding pot bearing.

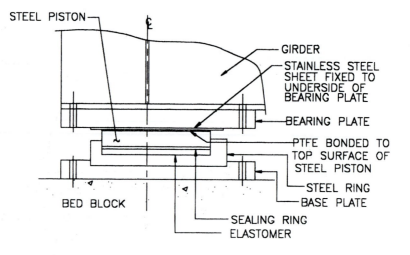

Fig. 4.6: Sliding pot bearing

4.4.6 Spherical Bearings

Spherical bearings are also capable of providing multidirectional rotations and horizontal movements at bridge supports and are very useful for skewed or curved bridge girders. Generally, a spherical bearing has two sliding surfaces—one at the matching spherical surface (similar to a ball and socket arrangement) which provides transverse and longitudinal rotations and another flat surface which provides horizontal movement. Both the sliding surfaces are required to be low friction surfaces and PTFE sheet mated against stainless steel may provide such condition.

In the foregoing paragraphs, the basic characteristics of a few bearing forms—both traditional and modern—have been discussed. There are other bearing forms such as disc bearings, seismic isolation bearings, etc., which employ fairly recent design concepts and incorporate or combine some of the characteristic behaviours of the bearings described earlier. However, research and development is an on going process and production of improved versions of bearings will continue. Therefore, if and when existing bearings of a bridge are required to be replaced, selection of the replacement bearings should not be restricted to the existing bearing forms only. A thorough search of the modern bearing forms with wider functional capabilities should be made. In fact, in some particular cases, modern bearings may be more suitable to overcome the problems which the existing traditional bearings failed to cope with.

4.5 PROBLEMS IN BEARINGS

As already noted, bridge bearings are required to serve two vital functions, viz. to transmit loads to the foundations and to allow movements of the supporting superstructure. It is therefore important that the bearings must be maintained properly, so that they can perform both these functions in the most satisfactory manner. However, being virtually the only components with movable parts in the bridge structure, bearings have to encounter many problems. Some of the more common problems are discussed in the following paragraphs.

4.5.1 Corrosion

As discussed in earlier chapters, corrosion in steel is caused by water coming in contact with steel surfaces. Presence of dust or debris on bearing surface substantially increases the chance of corrosion. In fact, accumulation of dust and debris is the most common cause of corrosion and rusting in bridge bearings. Debris have the tendency to absorb and retain moisture and other corrosive materials for a considerable period of time. Thus, if allowed to accumulate and remain on the surface, debris will lead to corrosion and rusting.

Corrosion effect is very pronounced in plate bearings, where the corrosion introduces irregularities in the unpainted sliding surfaces and increases the co-efficient of friction significantly. In roller bearings also, heavy corrosion on the contact surfaces may impede the movement of the rollers or even damage the shape of the rollers, rendering them incapable of any movements. Quite often, dirt accumulates on the rolling surface of the bearing plate and restricts the movement of the rollers. In rocker bearings, build up of corrosion on the contact surfaces of the rocker may also restrict the intended movement. Also, some of the smaller components, such as tooth bars in roller bearings (Fig. 4.3) or pins for fixed plate bearings (Fig. 4.2) suffer from the effect of corrosion, which may restrict the movement of the bearings or cause these components to shear off. Eventually the bearings get 'frozen' or jammed and become incapable of the required movements. This is likely to induce additional forces to the substructure. As a result the substructure which was not designed for these forces may show signs of distress, particularly in regions around the concrete bed block. This is a common occurrence in older bridges, where adequate reinforcement in the bed block was not always provided.

4.5.2 Mis-alignment

It is necessary that the bearings remain properly aligned during the entire life of the bridge, to enable these to allow longitudinal and/or rotational movements of the bridge with minimum resistance. Restriction to movements may cause additional forces in the structure, particularly in the substructure.

Mis-alignment of bridge bearings may be due to excessive and constant vibration or constant pounding of live loads, which may cause the bearing plates to move. Defective fabrication of bearings, error in setting the bearings (particularly for temperature) either at the time of construction or during subsequent maintenance work may also cause mis-alignment of bearings.

Bearings of a bridge with excessive skew are subjected to the effect of lateral rotation along with normal longitudinal rotation. This may also cause mis-alignment of the bearings, as traditional bearings (e.g. rocker, roller or plate bearings) are generally not designed to cater for such multidirectional rotations.

4.5.3 Tilting of Bearings

Tilting impedes the functioning of bearings apart from inducing eccentric load in the system. The eccentric load may cause overstressing of the bed plate and the concrete block underneath. In extreme case, the bed plate may fail in bending or the concrete at the edge of the bed plate be crushed due to the unintended eccentric load. Excessive tilting may even lead to the collapse of the bearing and induce additional forces in the structure.

One of the primary causes of tilting of the bearings is the movement of substructure. One reason for this is the unequal settlement of foundations. This may have been caused by inadequate design due, probably, to the lack of knowledge about sub-soil conditions at the time of construction. One other reason is the build up of water pressure behind the abutment due to inadequate provision for drainage of water, or due to clogging of existing weepholes. Tilting of bearings may also be caused by movement of the superstructure due to traction or braking forces induced by moving vehicles. Excessive and constant vibration on already distressed bearings may also cause tilting.

4.5.4 Damaged Anchor Bolts

Loose, bent or even sheared anchor bolts are the common problems associated with bearings in distress. There have been many instances where nuts have become loose and are even missing.

4.5.5 Damage in Bed Blocks

The concrete bed blocks on which the bed plates of bearings are fixed by anchor bolts are subject to substantial vertical and horizontal forces. In many of the older bridges the areas around the bed blocks have been found to have deteriorated. This deterioration may be attributed to various inter-related factors. Lack of proper contact between the underside of the bearing plate and the top of the bed block may lead to cracking and crushing of the bed block and masonry foundation underneath. Repeated impact loads from the vehicles due to poor roadway surface above the misaligned deck joints may aggravate the situation. Malfunctioning of bearings due to reasons discussed earlier in this section also adds to the distress. In the older bridges, the severity and the magnitude of these forces were often not anticipated and consequently not considered in the design. These under-reinforced (or unreinforced) bed blocks become easy victims of the unanticipated forces and show signs of distress.

4.5.6 Deterioration Due to Chemical Effects

Chemicals leaking from the vehicles on deck may make their way to the bearing surface and initiate the corrosion process. Presence of debris and dust on the bearings hastens the deterioration. The direct effect of corrosion has been dealt with in the preceding paragraphs. However, normally, leakage of chemicals from bridge deck is not a common phenomenon.

4.6 REMEDIAL SOLUTIONS

Whenever a bearing in a particular bridge is found to be in distress, a detailed study should be carried out to ascertain the cause of the distress, which should then be addressed for deciding on the corrective measures to be adopted. In this regard there are only two options open to the designer, viz. either rehabilitate the bearing or replace it. The decision either way will depend largely on the extent of the damage and the capability of the bearing to perform its required task. If the damage is of minor nature and the bearing can be made to perform its intended functions (viz. transmit the design forces and allow movements of the structure) by attending to the damaged components or re-setting or realignment, then replacement is not warranted and the existing bearing should be rehabilitated.

4.6.1 Rehabilitation

The term 'rehabilitation' normally means attending to the individual components which are in distress and if necessary to replace these to enable the bearing to perform its intended functions. Examples of these deficiencies are surface corrosion of non-contact surfaces, minor displacement of bearing plates or base plates, minor tilting of bearing, corrosion to anchor bolts, guide plates, tooth bars, etc.

Remedial solutions for some common problems encountered in the field are discussed in the following paragraphs:

(a) Corrosion

Where components do not show any major loss of section due to corrosion, these bearings may be cleaned *in situ* and greased. Normally bearings are to be painted except at the contact surfaces of rockers/rollers and sliding bearings, which should be oiled or greased. Bearings showing effects of severe corrosion may need temporary removal and thorough check up. For this purpose, the superstructure is to be jacked up to relieve the loads on the bearings and supported on temporary props. These should then be rehabilitated—if necessary, by replacing damaged components and re-erected after painting and greasing.

(b) Mis-alignment

A mis-aligned bearing can be rehabilitated similarly by first relieving the load from the bearing by jacking the superstructure and introducing temporary support system. The bearing components (roller, etc.) are then reset with correct alignment, giving due consideration to temperature effect for the inclination of the rollers.

As already noted, bearings of bridges with excessive skew are subjected to multi-directional rotations. Traditional bearings are not designed for such rotations and often suffer mis-alignment. Consequently, resetting of the existing bearings may not solve the mis-alignment problem permanently. In such cases, it is recommended to consider use of elastomeric or pot or spherical bearings, to accommodate this type of movement.

(c) Tilting of Bearings

In case of tilting of bearings, it will be advisable to investigate into most likely reason of the tilting. If it is found that the tilting is due to movement of the sub-structure, it will be necessary to first remove the cause of the sub-structure movement, such as cleaning clogged weep holes in the abutments, ensuring proper drainage behind the abutments, etc. Similarly, if the tilting of the bearing is due to movement of the superstructure, the remedial measures for such movements need to be implemented first. Resetting of tilted bearings without attending to the basic cause of the tilt will produce only a temporary solution and the problem may recur after some time.

Bearings suffering from tilting can be re-set in a similar manner, as done in the case of misaligned bearings, viz. jacking up the superstructure, introducing temporary props to support the superstructure and re-setting the bearing components including rollers with correct alignment and inclination for appropriate temperature. However, the base plates may need replacement due to change in the relative position of the new base plate *vis-a-vis* the rest of the bearing components, including the existing anchor bolts positions. Also, if required, the concrete bed-block should be repaired before placing the bearing.

4.6.2 Replacement

When a bearing shows major defects such as severe corrosion in rollers rendering them out of shape, cracks in the main components, etc. and is beyond repair or is likely to impede its functioning even after repair, it would be advisable to replace such a bearing in its entirety. In some cases, because of the remote situation of the bridge, it may be easier or even cheaper to replace a bearing rather than to repair it, thus avoiding the problems of locating and transporting skilled workmen as well as tools, tackles, etc. to work site. Even if it is found that some components could possibly be repaired at a workshop away from the bridge site, the cost of such repair work may be quite prohibitive and may warrant serious consideration for replacement of the bearing. Also, in such cases, the super-structure has to be kept resting on temporary support (in lieu of the bearing) till the repaired components arrive back at the site and the bearing is made operative again. Duration expected for such an operation is important, since temporary supports are normally designed with allowable overstresses for a short term use only.

Choice of the replacement bearing depends on many factors. Some of these are:

- The new bearing must satisfy the functional requirements, such as horizontal and/or rotational movements, as necessary.
- The overall height of the bearing has to fit in the existing available height. Otherwise major alteration in the existing structure may prove to be very expensive.
- Replacement of an existing free bearing made of steel by a new bearing with elastomer materials may lead to increased horizontal force to the substructure, due to possible higher shear resistance property of the elastomer material, compared to the frictional resistance of a steel bearing. Therefore the capacity of the sub-structure with the additional force needs to be checked.
- The new bearing should be compatible with the environment in which it will be used. As for example, in a moist environment an elastomeric bearing may be more suitable than steel bearing.

Selection of a replacement bearing should therefore be made with an open mind. Although there is bound to be a propensity to use the same type as the existing bearing, this should be resisted and a more appropriate new type of bearing may be considered for gainful use. The bearing should then be removed for providing access to the damaged bed block. After removal of the damaged concrete, the existing reinforcement should be cleaned, additional reinforcements added if necessary and new concrete of adequate strength cast. The bearing should be re-installed only after the concrete has hardened.

4.6.3 Anchor Bolts

When an anchor bolt is found bent or sheared off, it may be repaired by cutting off the portion above the top of the concrete and butt-welding a new threaded rod on top of the existing anchor bolt. If necessary the concrete around the anchor bolt may be removed locally. The welding should be tested to ensure full penetration of the weld metal. This method requires removal of the bearing temporarily for working on the anchor bolt.

4.6.4 Damaged Bed Blocks

Cracks and other defects in the concrete bed blocks should be investigated properly to establish the cause of the distress. If necessary major repair work or even replacement by new bedblock with stronger concrete may be carried out. For this purpose the superstructure should first be jacked up to relieve the loads and supported on temporary props.

BIBLIOGRAPHY

1. Mumber, J: 'Bearings', *Bridge Inspection and Rehabilitation*, Parsons Brinckerhoff, John Wiley & Sons Inc., USA, 1993.
2. Park, Sung H: *Bridge Rehabilitation and Replacement*, S.H. Park, Trenton, New Jersey, USA.

5

Connections and Selection of Fasteners

5.1 INTRODUCTION

Rehabilitation work of bridges generally involves strengthening of a few deficient members by addition of new materials or replacement of a number of members by new members of adequate strength. In some cases the existing structure may be sufficiently strong, but may need a few modifications at certain locations to satisfy present day functional requirement, such as increased head room or wider carriageway. In all such cases new connections are to be provided for securing new materials or members to the existing structure. Selection of fasteners and design of these connection details sometimes require lot of ingenuity and careful study. This is particularly so, when a new connection is located at or near an existing connection. In such a case, care should be taken not to disturb the behaviour pattern of the existing connection. Thus the characteristics of different fasteners and their behaviour pattern are likely to influence the selection and design of appropriate fastening arrangements for new connections. The present chapter deals with these aspects of different fastener systems.

5.2 TYPES OF FASTENERS

There are broadly three options for selection of fasteners for connections in the rehabilitation of existing bridges, viz. bolt, rivet and weld. Although progress in the use of bolting and welding during the recent decades have reduced the importance of riveting in structural steelwork, riveting has been successfully used in fabrication of steel bridges for many years. In fact majority of the steel bridges that we see today were constructed during the period when riveting was extensively used for bridge construction. Rehabilitation of such bridges necessarily requires working on the existing riveted connections. It is, therefore, necessary to understand the behaviour of riveted connections along with those of the alternative options such as bolts and welds for connecting new materials to the existing structure.

5.3 RIVETED CONNECTIONS

5.3.1 General

Riveting is essentially a forging process, in which a rivet (pre-formed with one head and shank) is inserted into a pre-aligned oversized hole on the elements to be connected, with the head tightly pressed against the elements. The shank of the rivet which protrudes through the hole to the other side is shaped into a driven head by pressure from a riveting machine. Figure 5.1 illustrates the process. Rivets used in structural steel fabrication are almost always heated to a minimum temperature of about 1000°C before they are inserted into the hole.

5.3.2 Assumptions in the Theory of Riveted Joints

In the traditional methods of analysis and design of riveted joints certain assumptions are made. Broadly, these are as follows:

(a) There is no frictional resistance between the members due to shrinkage of the rivet while cooling.
(b) Shear stress in the rivet is uniform over its entire cross-section.
(c) Bearing stress between the rivet and the member is uniform over their nominal contact surface.
(d) Bending of the rivet is neglected.
(e) Rivets in a group share direct (centric) loads equally.
(f) Tensile stress concentration near the rivet hole is neglected.

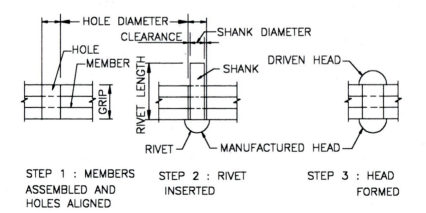

Fig. 5.1: Riveting process

(g) Rivets fill the holes completely.

Significance of these assumptions in the analysis and design of riveted joints is briefly discussed in the following paragraphs.

(a) Frictional Resistance

As the head of a rivet is being formed by pressure from a riveting machine, the hot rivet shank fills the clearance in the oversized hole. Subsequently, as the shank is allowed to cool slowly, it tends to shrink longitudinally as well as diametrically. The longitudinal shrinkage in the shank is resisted by the connected elements. As a result, tension (close to elastic limit) is developed in the shank, while the connected members are subjected to considerable compression. This compression produces frictional resistance to sliding in the interface of these members. The phenomenon of initial shrinkage of a rivet and the resultant friction between the connected elements make a riveted joint to behave in a way somewhat intermediate between a friction-type and a bearing-type connection. For design, however, the frictional resistance is ignored and a bearing-type connection is assumed.

(b) Shear Stress Distribution in a Rivet

Shear stress in a rivet is assumed to be uniform over its entire cross-section. However, under elastic conditions shear stress is not uniform. This incorrectness in the assumptions is taken care of by applying appropriate factor of safety while computing the permissible stress.

(c) Bearing Stress Distribution in a Rivet

Figure 5.2 (a) shows the radial nature of bearing stresses a rivet. For the purpose of design, however, these radial stresses are assumed to be equivalent to a uniform pressure (nominal stress) on a diametral plane through the rivet as shown in Fig. 5.2 (b)

(d) Bending of a Rivet

Once the friction between the members is overcome, theoretically a rivet is subjected to bending stress in addition to shear and bearing stresses. This

(a) ACTUAL BEARING STRESS IN RADIAL DIRECTION

(b) NOMINAL BEARING STRESS

Fig. 5.2: Bearing stress distribution in a rivet

bending may affect the distribution of bearing stress along the length of the rivet.

However, the effect of bending on a rivet is generally considered to be marginal, as the deformations are likely to be prevented by the friction between the members being connected.

(e) Sharing of Loads by Rivets in a Group

In traditional design methods, it is usually assumed that rivets in a group, when subjected to direct load will share the load equally. This concept is based on the assumption that the plates are perfectly rigid and the rivets perfectly elastic and is generally termed rigid plate method. According to this concept, the pure translation of one plate relative to the other produces equal deformation and unit shearing strain in all the rivets. Therefore, if the cross sections of the rivets are the same, the loads in each rivet would also be the same. However, in actual practice, the plates are not absolutely rigid and elongations in the plate length between the rivets are not the same. Therefore, the rivets do not share the loads equally. In fact, the rivets at the ends tend to carry more loads than those in the interior ones. This concept is usually termed elastic plate method. However, for design of rivet groups, the former (rigid plate) method is traditionally considered.

(f) Tensile Stress Concentration in the Member

Distribution of tensile stresses across a member containing one or more circular holes is not uniform within the elastic range. The stresses are dependent on the size and the location of the holes. Typical shapes of stress diagrams are shown in Fig. 5.3. The high stress concentration will cause the fibres nearest to the holes to reach yield points first. With increase of load these fibres will deform and stresses in the next fibres will increase. If this process continues, the distribution of stresses across the member will become more and more uniform until finally the load reaches the ultimate strength.

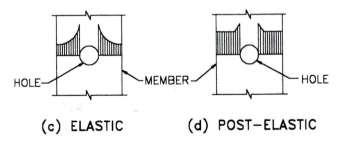

Fig. 5.3: Stress distribution in member across the hole of rivet

(g) Complete Filling of Rivet Holes

For the purpose of design, it is assumed that a heated rivet fills the over-sized holes completely during driving due to the high pressure from the driving machine or hammer. In actual practice, however, some rivets may fail to fill the holes and remain unattended inspite of inspection system. These rivets receive full load only after the other rivets in the group become fully loaded and a re-distribution of forces takes place.

5.3.3 Behaviour of Rivets in Joint

The most common mode of load transmission in a riveted joint is by shear. The load is transmitted through bearing between the surfaces of the plate and the shank producing shear in the shank. In this case, the shank may have one, two or more shear planes as shown in Figs. 5.4 (a), (b) and (c) respectively. Figure 5.4 (d) shows a case where the load is transmitted through bearing between the surfaces of the plate and the rivet head, producing tension in the shank.

5.3.4 Failure Modes of a Riveted Joint

A riveted joint normally deforms appreciably before failure. Failure of a joint may take place in any one of the following modes, depending upon which one is the weakest. Different modes of failure have been illustrated in Fig. 5.5.

(a) Shear Failure of a Rivet

Shear failure of a rivet occurs across one or more planes between the members it connects, depending on whether the rivet is in single shear or in multiple shear.

(b) Bearing or Crushing Failure of a Rivet

This mode of failure occurs at the half-circumference contact surface of the rivet and the member. This may occur due to bearing failure of the member or of the rivet or of both.

(c) Tension Failure or Tearing of the Member

This mode of failure may occur when the cross-sectional area of the member is inadequate to transmit the load.

(d) Shear-out Failure of the Member

This failure occurs when there is insufficient edge distance in the member along line of the load. This type of failure can be avoided by providing adequate edge distance, as stipulated in various standards or codes.

(a) LAP JOINT : RIVET IN SINGLE SHEAR

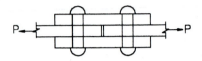
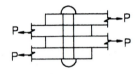
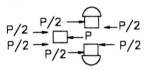
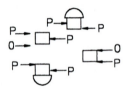

(b) BUTT JOINT : RIVET IN DOUBLE SHEAR

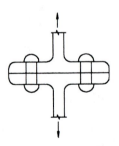

(c) RIVET IN MULTIPLE SHEAR

(d) RIVETS IN TENSION

Fig. 5.4: Behaviour of rivets in joint

5.4 BOLTED CONNECTIONS

5.4.1 General

Bolting is a simple and sufficiently reliable method for connecting steel elements. It does not require any particularly sophisticated equipment for

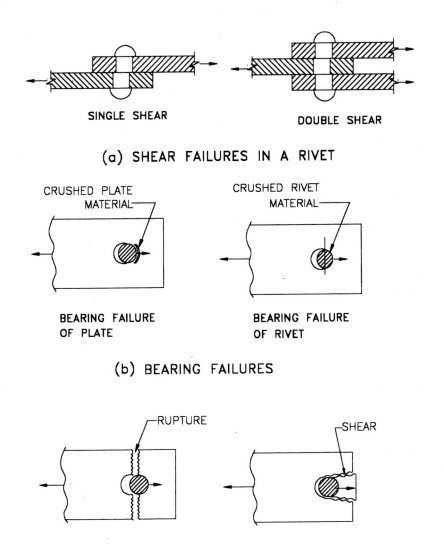

SINGLE SHEAR

DOUBLE SHEAR

(a) SHEAR FAILURES IN A RIVET

CRUSHED PLATE MATERIAL

CRUSHED RIVET MATERIAL

BEARING FAILURE OF PLATE

BEARING FAILURE OF RIVET

(b) BEARING FAILURES

RUPTURE

SHEAR

(c) TENSION FAILURE OR TEARING IN A PLATE

(d) SHEAR OUT FAILURE IN A PLATE

Fig. 5.5: Modes of failure in a riveted joint

installation and is generally quicker to install compared to other types of fasteners. Because of these advantages, use of bolting has become very widespread in field operations.

5.4.2 Types of Bolts

Broadly, there are three types of bolts which are used in a bolted joint:

(a) Unfinished Bolts

These bolts (sometimes called 'black bolts') are forged from rolled steel round bars. These have large dimensional tolerances on shank and thread and are used in "clearance" hole with diameters normally 1.5 mm greater than the diameter of the shank. These bolts are generally manufactured from mild steel bars, but can also be of high strength steel.

(b) Close Tolerance Bolts

These bolts are made from rolled steel bars and turned down to closer tolerances (+0.000 mm, − 0.125 mm) and are used in close tolerance holes (+0.125, -0.000) generally produced by reaming. These bolts are inserted into the holes by means of light blows of a hammer. The faces under the head and nut are usually machined; washers to be used are also machined on both faces. It is customary to specify the length of the thread in order to avoid threads encroaching into the steelwork connected. These bolts are generally manufactured from high strength steel bars.

In some close tolerance bolts the threaded portion is of lesser diameter than the shank or barrel diameter. This variation eases the site operation considerably, as there is less chance of the thread getting damaged during insertion of the bolt in the hole. These bolts are generally known as turned barrel bolts.

(c) High Strength Friction Grip Bolts

Comparatively a recent development, these are high strength bolts with high strength nuts and hardened steel washers. These can be tightened to give a high shank tension, which clamps the joining members between the bolt head and the nut. Load is transmitted by the friction developed between the interfaces of the members, and not by shear or bearing as in the case of other types of structural bolts.

5.4.3 Behaviour of Bolts in Joints

The modes of transmission of load in a bolted joint are somewhat similar to those in a riveted joint, viz. shear, bearing, tension or friction. However, the actual manner of load transmission in a bolted joint varies significantly from that in a riveted joint, depending on the type of bolts used in a joint.

In a joint with unfinished (black) bolt, the load is considered to be transmitted by bearing and shear, ignoring the friction between the interface of the members due to tightening of the bolt. When the load is applied, the members slip until the shank comes in contact with the edge of the

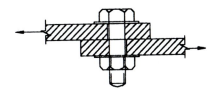

Fig. 5.6: Joint with unfinished bolts

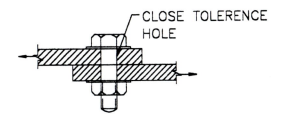

Fig. 5.7: Joint with close tolerance or turned barrel bolt

hole (Fig. 5.6) and the load is then transmitted by bearing and shear. The shear stress across the cross section of the shank is assumed to be uniformly distributed. However, the bearing between the bolt and the member is concentrated on the edge of the member and is likely to produce high local stresses, until plastic deformation takes place. In any case, in locations of main connections in bridges, where reversal of forces may occur and slippage is undesirable, these bolts with clearance holes are not recommended to be used.

When members are joined by a close tolerance or a turned barrel bolt fitted in a reamed close tolerance hole, the slip in the joint is minimal and the load is transmitted directly by shear and bearing (Fig. 5.7) as in the case of a riveted joint. The high local bearing stresses associated with bolts in clearance holes are also minimised. Behaviour of a joint with close tolerance or turned barrel bolts is thus considered to be much better than that of a joint with unfinished bolts.

When a bolt is subjected to tension, the critical section is at the root of the thread where the area is least. At this section there is also a high stress concentration resulting in a high local stress. If the load is increased the material at the root yields and the stress distribution becomes more uniform. Therefore, the ultimate strength of the bolt is not affected by the stress concentration.

Bending in short bolts is normally ignored in the design. However, in bolts with grips exceeding five times the diameter, the effect of bending should be checked.

The mechanism of a friction grip bolted joint is illustrated in Fig. 5.8. As the nut is tightened to produce a tensile force T in the shank of the bolt approaching yield point, the connecting members are clamped tightly together, developing a frictional resistance F. The maximum value of F is given by

$$F = \mu T$$

where μ is the co-efficient of friction between the members at the interface. The value of T is taken as 0.8 or 0.9 of the yield strength of the bolt. If the external load P does not exceed F, the joint will transmit the load without any slip.

As the behaviour of a friction grip bolted joint depends on the tension produced in the shanks of the bolts, it is imperative that the bolts are tightened to the required tension, as otherwise slip may occur at service load condition, rendering these bolts to behave as ordinary unfinished (black) bolts. There are a number of methods to ensure correct shank tension in the field condition, such as part-twist method, calibrated-wrench method, load indicating washers, or twist-off bolts.

Frictional resistance in a friction grip joint depends not only on the adequate tightening of the bolts, but also on the surface condition at the interface of the members connected. It is, therefore, necessary that these surfaces are free of defects, which would prevent proper seating of the parts, especially dirt, barr and similar other foreign material. Also the interface within slip critical joints should be free from oil, paint, lacquer, galvanising or similar finishes, which might prevent the development of friction.

One other point of caution during installation of a large friction grip joint is that there is always a danger of bedding down of the connected piles; as a result, tension in some of the bolts tightened first may have been

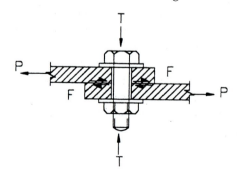

Fig. 5.8: Joint with friction grip bolt

reduced by the time the latter bolts are tightened. This danger needs to be guarded against for any large group of friction grip bolted joint.

5.5 WELDED CONNECTIONS

5.5.1 General

Welding is a comparatively modern process of joining together metal parts by application of heat resulting in fusion of the two sections along the line of the joint. The most common type of welding process used in structural steelwork is the electric arc process. In this process an intense heat source from an electric arc operating in the gap between the tip of an electrode and the steel components to be welded together is used to melt and fuse the parent metal of the components along with the metal of the electrode. This, on cooling, forms a solid weld that joins the two metal parts into one piece.

The two main systems of the electric arc process are:

(a) Manual Arc Welding
In this system the electrode is in the form of a hand-held metal stick coated with flux, which melts and protects the molten metal from oxidation and improves the mechanical properties of the welds.

(b) Automatic Arc Welding
In this system a continuous wire or electrode is automatically fed into the welding zone by unrolling a spool of the wire or electrode. The wire may be coated with flux. Alternatively, the flux may be separately poured from a hopper directly over the end of the electrode, covering the welding zone and isolating it from air. This process normally produces a homogeneous weld with deep penetration, possessing high mechanical properties. In a variation of this process, termed, 'semi-automatic welding', the wire is fed mechanically while the movement along the weld is done manually.

The automatic and semi-automatic processes are generally used in the workshops. The manual arc welding process, on the other hand, is very useful and handy for use in the workshops as well as at the work-sites.

5.5.2 Types of Welds

The two types of welds in common use are (a) fillet weld and (b) butt weld

(a) Fillet Weld
In fillet weld the weld metal is deposited outside the profile of the joining elements (Fig. 5.9). Although the size of fillet weld is specified by the leg

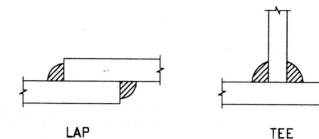

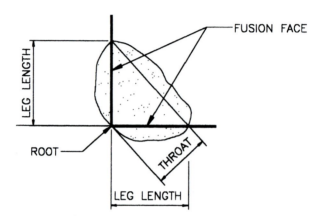

Fig. 5.9: Fillet weld

length, the throat thickness is considered for computing its strength. In a fillet welded joint, the layout of the weld is important since stress distribution in a fillet welded joint depends not only on the load but also on the layout of the weld or group of welds.

(b) Butt Weld

In a butt welded joint the edges of the members are butted against each other and joined by fusing the metal to produce a continuous joint. Depending upon the current used, arc can melt the metal up to a certain depth only. If the thickness of the members to be joined by butt welding is more than this depth, the edges of the members are required to be 'prepared' to form a groove along the joint-line, so that joint continuity through the thickness can be achieved. The prepared groove is then filled by weld metal from the electrode. Figure 5.10 shows a few common types of butt welds.

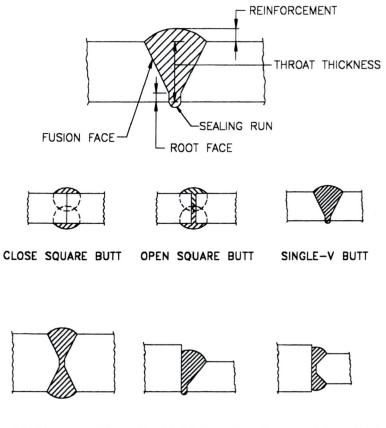

Fig. 5.10: Typical butt welds

Selection of the electrode and edge preparation as also the properties of the parent metals play vital roles in developing the strength of such a joint.

The defects which are commonly noticed in welds are shown in Fig. 5.11

5.5.3 Behaviour of Welded Joints

The following paragraphs briefly deal with some of the important aspects which influence the behaviour of welded joints.

(a) Characteristics of Welds

The solidified weld metal which joins the components is essentially a mixture of the molten parent metal and the electrode, and is usually made

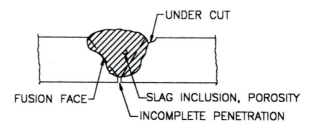

Fig. 5.11: Defects in welds

stronger than the components by judicious adoption of the composition of the electrode.

As the molten metal cools and solidifies, most of the heat flows into the adjoining parent metal. This causes metallurgical changes in the structure of the steel up to a certain distance, depending on the maximum temperature attained during the welding process. This region is called heat affected zone (HAZ). The changes in the structure in this zone depends on the chemical composition of the steel, particularly on its carbon content and the rate of cooling. Slow cooling usually makes the steel ductile, whereas rapid cooling makes it brittle. In turn, the cooling rate is dependent on a number of factors, viz. heat input from the arc, thickness of the components, type of joint, temperature of the components prior to welding, etc. In a thick material, the rate of cooling is faster than in a thin one. It is a common practice to pre-heat thick components. Pre-heating reduces the temperature gradient between the weld and the adjoining metal and helps to reduce brittleness. Thus the consequent risk of cracking in the region is largely minimised.

(b) Composition of Steel

Carbon is the primary strengthening element in steel. However, increased carbon content may impair ductility and weldability of the steel. Therefore, in order to obtain enhanced properties in the steel (while keeping carbon level very low), other admixtures or alloys are generally added during the process of steel making.

The relative influence of chemical contents on the weldability of a particular steel is guided by the value of 'Carbon Equivalent' which is derived from the following empirical formula:

$$CE = C + (Mn/6) + (Cr + Mo + V)/5 + (Ni + Cu)/15$$

where the chemical symbols represent the percentage of the respective elements in the steel. The higher the CE, the lower will be the allowable cooling rate. Consequently, harder and more brittle will be the heat affected zone (HAZ) and more susceptible will it be to cracking. Therefore, with the increased value of CE, use of low hydrogen electrodes and pre-heating becomes important.

(c) Fatigue

As discussed earlier, a structural member subjected to cyclic loading may fail through initiation and propagation of cracks at stress levels much lower than those required for failure under steady static loading. This phenomenon is known as fatigue.

In bridges with welded details, these cracks are most likely to start from welds rather than from any other location. Generally, structural welds have a rough profile with minute metallurgical discontinuities, from which cracks may start. Cracks are initiated at notchlike details where high concentration of stress develops. Toes and roots of welds are common examples of such notches (Fig. 5.12). Also a welding arc strike that does not deposit weld metal on the parent member, may embrittle the metal due to fast cooling rate and initiate minute cracks.

A fatigue crack is generally difficult to detect at the initial stage and is likely to escape notice during inspection at this stage. However, once the crack becomes significant, it progresses rapidly and needs to be attended to almost immediately.

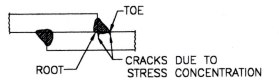

Fig. 5.12: Local stress concentration in welds

(d) Shrinkage, Residual Stress and Distortion

The hot metal in the weld zone shrinks during cooling causing residual stress and possible distortion in the joint. The distortion may be transverse or longitudinal. Figure 5.13 shows a few examples of such distortions.

For controlling distortions the following points are worth consideration:
- Shrinkage is proportional to the volume of weld metal. Therefore, weld size should be kept to the minimum while designing.
- As far as possible, the arrangement or disposition of welds should be symmetrical.
- Welding procedure (i.e., choice of electrode, edge preparation, preheat, voltage, current, travel speed, welding position, number of runs, sequence of welding, etc.) should be judiciously selected.

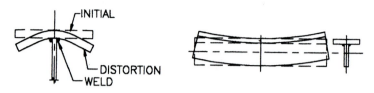

(a) TRANSVERSE DISTORTION (c) LONGITUDINAL DISTORTION

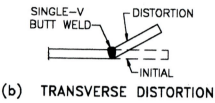

(b)   TRANSVERSE DISTORTION

Fig. 5.13: Shrinkage and distortion in a welded joint

(e) Welding Position

Figure 5.14 shows some common welding positions. It is obvious that welding in down-hand position is much easier to perform than welding in overhead position, which requires greater skill to produce good results and also consumes more time. Welding in overhead position should, therefore, be avoided as far as possible.

(f) Access Requirements

Proper access is essential for a good weld. This is particularly significant when a member is to be welded to an existing structure at a location with hardly any room for positioning the electrode. This aspect needs special consideration while finalising the connection details.

From the foregoing, the primary factors which determine the strength and durability of a welded connection may be summarised as follows:

- Composition of steel
- Choice of electrode
- Welding parameters (voltage, current and travel speed)
- Edge preparation
- Joint detail/number of runs
- Sequence of welding
- Rate of cooling/preheat

- Welding position
- Proper access

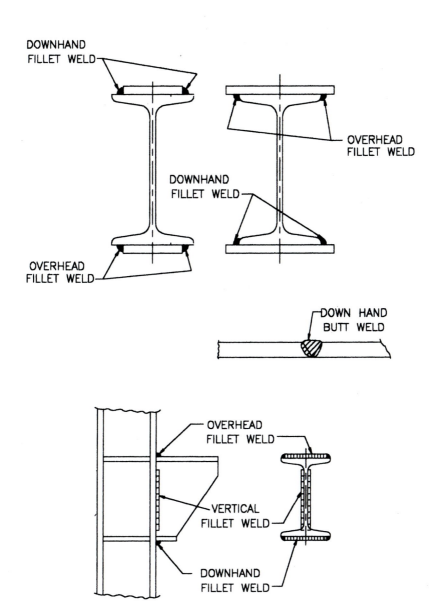

Fig. 5.14: Welding positions

5.6 OTHER FACTORS INFLUENCING SELECTION OF ARRANGEMENT OF FASTENERS

Apart from the characteristics of the fasteners and their behaviour in joints, there are some other important factors which also contribute to the final selection of a fastening arrangement. These are briefly discussed below:

5.6.1 Adequacy

A member is as strong as its weakest link. Therefore, the connection details must be adequate to match the design forces, so that the rehabilitated structure can behave in the manner intended by the designer.

5.6.2 Skilled Labour

In general rehabilitation work is labour intensive. Availability of labour or of a particular type of skilled labour may be of vital importance. This aspect assumes greater significance in remote areas, where labour skilled in a particular trade (e.g., machining, welding, etc.) may not be available. Importing such labour force may prove to be impractical and costly.

5.6.3 Equipment

Availability of equipment should be of prime consideration in the development of a rehabilitation scheme. At the initial stage it should be considered whether it will be feasible or not to mobilise a particular equipment and the scheme should be developed accordingly.

5.6.4 Raw Material

Availability of raw material and consumables which are required for the proposed connection should be ascertained before finalising the details.

5.6.5 Ease of Site Operation

Designers should always recognize the practical considerations of the erection process at site. Simplicity of detail, ease in handling, access to the work, practicability of scaffolding etc., are some of the important factors for consideration. Ease of approach for inspection during rehabilitation and subsequent maintenance should also be given high priority by the designer.

5.6.6 Duration of Work at Site

Rehabilitation work is almost invariably carried out with minimum interruption to traffic flow by imposing traffic restrictions for short intervals or by other means. Installation of new members and completion of connection to existing ones are thus required to be done at the shortest possible time. This factor may play a significant role in the selection of the fastener system.

5.7 CONCLUDING REMARKS

Many of the existing steel bridges are of riveted construction. Rehabilitation or repair work on these bridges is likely to involve working on the existing riveted connections for fixing new members. In such cases a suitable detail utilising the existing rivet holes is prepared for fixing the new members. The obvious choice would be to remove the existing rivet and fix the new (pre-drilled) members by driving new rivets at site. However, in many countries, riveting is fast becoming (if not already become) obsolete. Also, it may not be easy to arrange power, riveting equipment and skilled labour group at a distant work site. Bolted connections in such cases will be the most preferred alternative. Bolts in clearance holes are not suitable for use in bridges where reversal of forces may occur and slippage is undersirable. Consequently, a slip-critical bolted connection is required to be provided. Close tolerance bolts of the size of the rivet holes may be most suitable as alternative to rivets. These are basically 'slip-free' bearing type connections. like rivets and transmit loads by bearing and shear. Also, as discussed earlier, although a rivet is not designed to transmit load by frictional resistance, it does develop frictional resistance at the interface of the connecting members. For achieving this behavioural similarity with adjoining rivets, high strength (in preference to mild steel) bolt is recommended. In a high strength bolt adequate tension in the shank can be achieved while tightening the nut, thereby developing frictional resistance at the interface of the members. The existing level of tension and frictional resistance are unlikely to be achieved by mild steel bolts, as the tension area at the core of the threaded portion of the bolt is less than that of a rivet. Hence the preference for high strength bolt.

Although a bolt is costlier than a rivet, bolted connections are invariably more attractive, because these are fast to install, less noisy and do not require particularly special equipment or highly skilled and experienced workmen. These are very useful when limited number of fasteners are required to be installed and that too at different locations spread all over the bridge structure.

Use of friction grip bolts instead of close tolerance bolts may seem to be an attractive alternative for work on an existing riveted connection. However, there is a danger that on tightening a friction grip bolt by torque wrench or similar device, some of the existing adjoining rivets may become loose due to bedding down of the connected piles. This aspect needs careful consideration for work on an existing riveted connection. Where a new connection does not involve existing riveted joint, either close tolerance bolts or friction grip bolts may be conveniently used.

Because of flexibility of design and comparatively easy and fast site operation, welding often appears to be an economic and attractive alternative to bolting. However, welding is a comparatively sophisticated process and the strength and durability of a welded joint depend on many factors including the composition of the parent steel. It is, therefore, imperative that the properties of the existing steel, the effect of welding on its fatigue strength and other related parameters are duly examined before a final decision is taken. The process is no doubt time consuming and fraught with uncertainties. Therefore, it may perhaps be advantageous to avoid welding on old bridges and explore solutions with conventional fasteners such as bolts.

BIBLIOGRAPHY

1. Dowling, PJ, Knowles, PR and Owens, GW (Ed): *Structural Steel Design*, The Steel Construction Institute, 1988.
2. Gaylord, EH Jr and Gaylord CN: *Design of Steel Structures*, McGraw Hill Book Co. Inc., New York, 1957.
3. Brester, B and Lin TY: *Design of Steel Structures*, John Wiley & Sons, Inc. New York 1960.
4. Grinter, LE: *Design of Modern Steel Structures*, The Macmillan Co, New York 1965.
5. Brockenbrugh, RL and Merrit FS (Ed.): *Structural Steel Designer's Handbook*, McGraw-Hill Book Co. Inc., New York, 1994.
6. Tordoff, D: *Steel Bridges*, The British Constructional Steelwork Association Ltd, London, 1985.
7. Needham, FH: 'Connections in Structural Steelwork', *The Structural Engineer,* September 1980, The Institution of Structural Engineers, London.

6

Analysis of Typical Connections

This chapter presents the basic principles of analysis for a few typical connections which are commonly used in rehabilitation and repair work of steel bridges.

6.1 BOLTED CONNECTIONS

6.1.1 Concentric Shear Connection

Figure 6.1(a) shows a concentrically loaded shear connection in a butt joint where two tie plates are spliced by two cover plates placed on either side of the connecting plates and secured by bolts of same diameter.

The layout of the connecting bolts is such that the centre of area of the group of bolts on either side of the joint lies in the line of the forces. This is defined as a direct or concentrically loaded connection. As the plates are assumed to be rigid, the force can be considered to be shared equally by the bolts in a group. Thus the force F in each bolt would be:

$$F = P/n$$

where P is the force to be transmitted and n is the number of bolts in a group.

In an unusual case where bolts of different diameters are used in the same group, it can be assumed that the force in a particular bolt would be proportional to the total area of the bolts in the group. In order that the connection can be considered as a concentrically loaded one, the line of force must coincide with the centre of area of such a bolt group. For computing the centre of area of such a group the differing areas of the bolts are to be duly considered.

Figure 6.1(b) shows another example of direct joint where two members are connected by lap joints. In this case the analysis would be similar to the one discussed in the previous paragraphs.

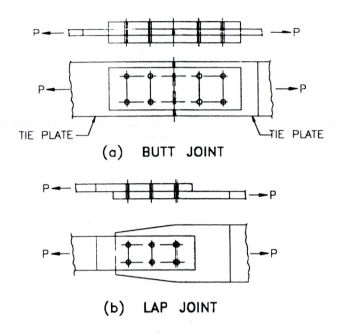

(a)   BUTT   JOINT

(b)   LAP   JOINT

Fig. 6.1: Concentric shear connections

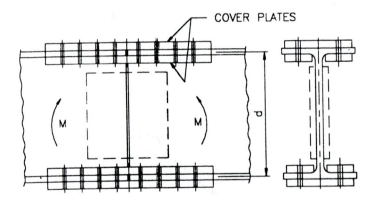

Fig. 6.2: Flange splice

Figure 6.2 shows the details of splice in the flanges of a girder where the flange forces (due to moment) are transmitted by cover plates secured to the flanges by bolts. If the moment at the section is M and d is the distance between the centres of the flanges, the force in the flange is M/d. The geometry of the connection is such that the splice can be analysed as a concentric shear connection. Analysis for web splice to transmit the shear has been discussed in the next section. Contributions for web fasteners to bending resistance is generally ignored, except in the case of deep girders.

6.1.2 Eccentric Shear Connection

Figure 6.3(a) shows a bracket connection in which a load P is applied in the plane of the connection at an eccentricity e from the centroid 'G' of the bolt group. This eccentric load is equivalent to a concentric force P passing through 'G' and a moment $P.e$ (Fig. 6.3b) which tends to rotate the side plates about 'G'. For the purpose of analysis, these two load conditions may be treated separately and then the results can be superimposed to get the combined effect.

(a) Concentric Force

As the plates are assumed to be rigid and the line of force P passes through 'G', the centroid of the bolt group, the force will be shared equally by each bolt. Thus, if there are n number of bolts in the group, the shear on each bolt is $F_a = P/n$.

(b) Moment

To calculate the force on a bolt due to moment, it is assumed that the shearing force F_m on any bolt (due to the moment) is proportional to its distance from the centroid of the bolt group 'G' and that the force F_m acts normal to the line joining the concerned bolt and 'G'. Thus, the bolt farthest from the centroid of the bolt group will carry maximum load.

Consider the group of bolts shown in Fig. 6.3(c). The bolts B_1, B_2 etc., are at a distance r_1, r_2 etc. from the centroid 'G' of the group. The force on the bolt B_1, which is farthest from 'G' is say F_{m1}. In consideration of the assumption that force in a bolt is proportional to its distance from 'G', it follows that force on B_2 (located at a distance r_2 from 'G') will be $F_{m1} \times r_2/r_1$. Similar will be the forces for other bolts also.

Moment of resistance of the bolt group will be:

$$MR = F_{m1}\, r_1 + F_{m1} \cdot \frac{r_2}{r_1} \cdot r_2 + F_{m1} \cdot \frac{r_3}{r_1} \cdot r_3 + \ldots$$

$$= \frac{F_{m1}}{r_1} \left[r_1^2 + r_2^2 + r_3^2 + \ldots \right]$$

$$= \frac{F_{m1}}{r_1} \cdot \sum r^2$$

Equating this to the applied moment,

$$P.e = \frac{F_{m1} \sum r^2}{r_1} \quad \text{or} \quad F_{m1} = \frac{P.e.r_1}{\sum r^2}$$

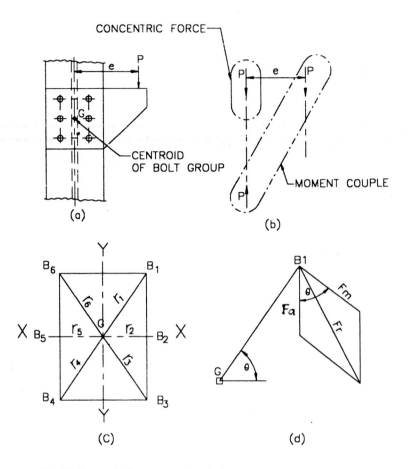

Fig. 6.3: Eccentric shear connection (bolt group in shear and torsion)

F_{m1} is the force on the most heavily loaded bolt B_1

(c) Combined Effect

The two forces F_a due to direct load and F_m due to moment for a bolt have different directions. The resultant force F_r can be found graphically as shown in Fig. 6.3(d). The algebraic formula can be derived as follows:

$$F_r = \sqrt{(F_a)^2 + (F_m)^2 + 2 \cdot F_a \cdot F_m \cdot \cos \theta}$$

where θ is the angle between the forces F_a and F_m.

A typical web splice is shown in Fig. 6.4. This is a common example of eccentric shear connection, which in effect consists of two eccentric connections with the common shear force *V*. Analysis of the bolt groups can be done in the same manner as has been shown for a bracket in the previous paragraphs.

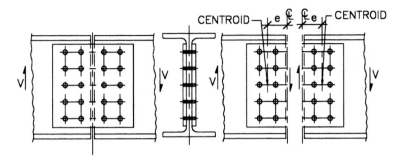

Fig. 6.4: Web splice

Another common example for eccentric shear connection is found in end-cleated connections of girders in bridge deck system. There are mainly two possible behaviours of the connection which would affect the analysis, viz, the main girder is either free to rotate or is prevented from rotation.

As an example of the first possibility, Fig. 6.5(a) shows a one-sided connection of a secondary girder to a main girder with low torsional rigidity, which is assumed to rotate with the application of shear load from the secondary girder. The shear *V* is considered to be transmitted to the web of the main girder by means of double angle cleats. The bolt group connecting these cleats to the secondary girder is to be analysed for eccentric moment due to the shear *V* acting at a distance *e* from the centroid of the bolt group as shown in the figure. The bolts connecting the cleats to the web of the main girder are required to resist the vertical shear force *V* only.

A variation to the previous example is shown in Fig. 6.5(b) where two secondary girders from two sides are connected to the main girder. With this double sided connection detail, the main girder is prevented from rotation. If the cleats are sufficiently stiff and do not rotate under the applied shear load from the secondary girder, the bolt group connecting the cleat to the web of the main girder will be required to resist not only the vertical shear, but also tension due to moment *V.e*. Analysis of moment connection with bolts in shear and tension has been discussed in the following paragraphs.

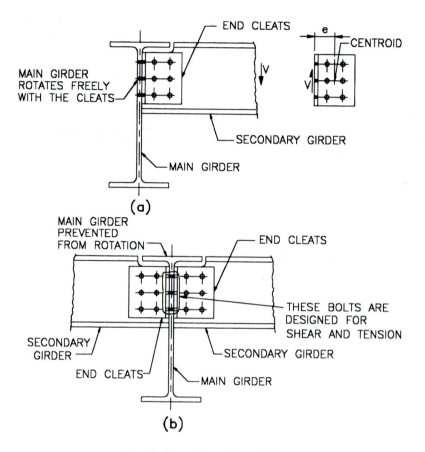

Fig. 6.5: Connections with end cleats

6.1.3 Connections with Bolts in Shear and Tension

Figure 6.6 illustrates a few cases of connections subjected to combined shear and tension. In these cases the applied moment is not in the plane of the bolts and tends to rotate the joint across the plane of the bolts, thereby inducing tension in the bolts in addition to shear. The conservative approach for analysis is to consider the shear to be shared equally by all the bolts and check the top-most bolts for combined shear and tension. These connections can be treated somewhat similar to a typical bracket connection shown in Fig. 6.7(a), in which a load P is applied at an eccentricity e from the edge of the bracket. In a simplified method for analysis the centre of rotation is assumed to be at the bottom bolt of the group, the loads varying linearly as shown in the figure.

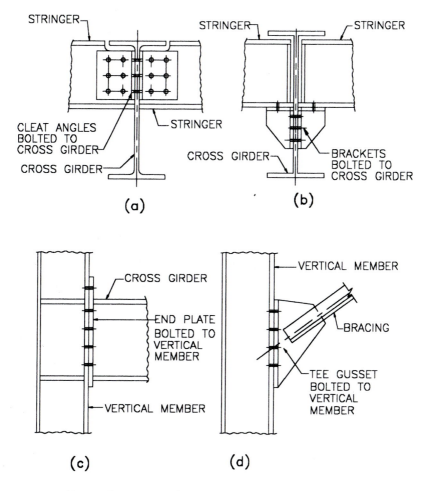

Fig. 6.6: Connections subjected to combined shear and tension

If T_1, T_2, T_3, etc. be the tensions in bolts 1, 2, 3, etc. located at distance r_1, r_2, r_3 etc. from the bottom most bolt, by similar triangles, then

$$\frac{T_1}{r_1} = \frac{T_2}{r_2} = \frac{T_3}{r_3} \quad \text{etc.}$$

If M_1, M_2, M_3, etc. are moments shared by the bolts marked 1, 2, 3, etc. respectively, considering two rows of bolts the total moment resisted by the bolt group is:

$$M = 2(M_1 + M_2 + M_3 + ...)$$

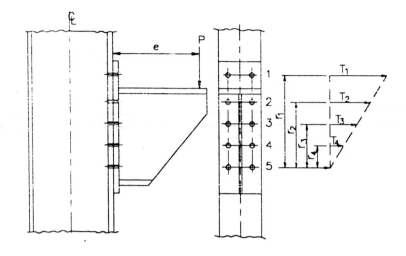

(a)   BRACKET CONNECTION

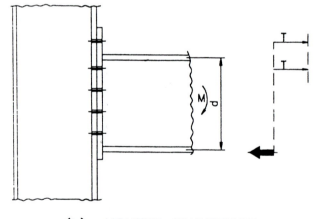

(b)   MOMENT CONNECTION
(ALTERNATIVE METHOD)

Fig. 6.7: Bolts in shear and tension

$$= 2\,(T_1\,r_1 \,+\, T_2\,r_2 \,+\, T_3\,r_3 \,+\, ...)$$

$$= 2\,\frac{T_1}{r_1}\,(r_1^2 \,+\, r_2^2 \,+\, r_3^2 \,+\, ...)$$

$$= \frac{2 T_1 \sum r^2}{r_1} = P.e$$

The maximum bolt tension on the top most bolt is

$$T_1 = P.e.r_1/2 \Sigma r^2$$

The load F_s due to direct shear is

$$F_s = P/\text{No. of bolts}$$

The top most bolt is to be checked for combined shear and tension.

In an alternative method, only the bolts located at the top flange are considered to take the tension along with their share of the shear.

Referring to Fig. 6.7(b) tension is top group of bolts is $n.T = M/d$, where M is the applied moment, d the distance between the centres of flanges and n the number of bolts considered as tension bolts. These bolts are to be checked for combined shear and tension.

6.2 FILLET-WELDED ECCENTRIC CONNECTIONS

Two types of eccentric connections are considered here:
(a) Load lying in the plane of weld
(b) Load not lying in the plane of weld
    In both these types the fillet welds are subjected to shear due to direct load and moment.

6.2.1 Load Lying in the Plane of Welds

The theory for analysis of eccentric shear connections for bolt groups can be broadly applied in the welded joint also (Fig. 6.8). While in a bolt group, bolts are discrete individual connections, in a welded connection the weld is a continuous connection and the analysis has to be done accordingly. The connected members are assumed to be rigid. The eccentric load P is applied at an eccentricity e from the centroid 'G' of the weld. This eccentric load is equivalent to a concentric force P passing through 'G' and a moment $P.e$, which tends to rotate the side plates about 'G'. These two load conditions are treated separately and then the results are superimposed to get the combined effect. The shear due to the concentric force P is assumed to be uniform throughout the weld. The force in the weld due to the moment is considered to be directly proportional to the distance from the centroid of the weld. Thus the weld farthest from 'G' will carry maximum load. The combined effect of direct load and moment can be obtained by adding the two results vectorially.

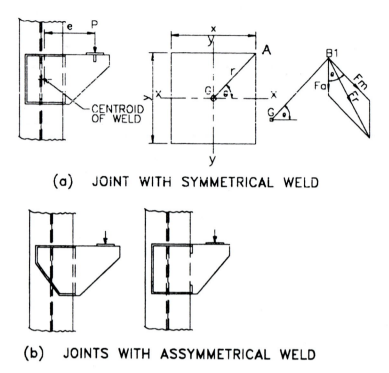

(a)   JOINT WITH SYMMETRICAL WELD

(b)   JOINTS WITH ASSYMMETRICAL WELD

Fig. 6.8: Eccentric welded connections (load lying in the plane of weld)

Assuming the weld to be of unit leg length and uniform throughout, the shear force in the weld due to the concentric force P is given by

$$F_a = P/L$$

where L is the total effective length of weld. Torsional moment $P.e$ produces bending forces F_m in the weld about an axis passing through 'G' and perpendicular to the plane of the weld and is given by:

$$F_m = \frac{P.e.r.}{I_p}$$

where r is the distance of a point under consideration from 'G' and I_p is the polar moment of inertia of the weld and is given by:

$$I_p = I_{xx} + I_{yy}$$

$$I_{xx} = 2\frac{y^3}{12} + 2x\left(\frac{y}{2}\right)^2$$

$$I_{yy} = 2\frac{x^3}{12} + 2y \left(\frac{x}{2}\right)^2$$

where x and y are the lengths of weld along X- and Y-axes

In Fig. 6.8(a), 'A' is a point farthest from the centroid of the weld and is the heaviest loaded weld. The distance r of 'A' and 'G' is given by:

$$r = \sqrt{\left(\frac{x}{2}\right)^2 + \left(\frac{y}{2}\right)^2} = \frac{1}{2}\sqrt{(x^2 + y^2)}$$

The resultant force F_r at 'A', the position of maximum shear, is given by:

$$F_r = \sqrt{(F_a)^2 + (F_m)^2 + 2 \cdot F_a \cdot F_m \cdot \cos \theta}$$

where θ is the angle between AG and XX axis.

If the weld layout is not symmetrical as in Fig. 6.8(b) the centroid of the weld is to be found first to determine the values of eccentricity and the polar moment of inertia. The stresses can be obtained as above.

6.2.2 Load not Lying in the Plane of Welds

A simple bracket connection is shown in Fig. 6.9. In this case also the eccentric load P, applied at a distance from the plane of the weld, can be considered equivalent to a direct force P, passing through the weld and a moment $P.e$ which tends to rotate the joint across the plane of the weld. The load on the weld can be obtained by applying beam bending formula.

Assuming the weld to be of unit leg length and uniform throughout, the shear force in the weld due to the direct force P is given by:

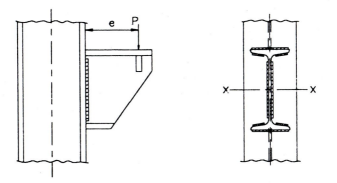

Fig. 6.9: Eccentric welded connections (load not lying in the plane of weld)

$$F_a = P / \text{length of weld}$$

The load due to moment is given by

$$F_m = P.e.y/I$$

where I is the moment of inertia of the weld and y is the distance of the farthest weld from the neutral axis of the weld layout.

The resultant force F_r is given by:

$$F_r = \sqrt{(F_a)^2 + (F_m)^2}$$

Typical Remedial Solutions

7.1 INTRODUCTION

This chapter deals with some typical problems encountered on existing steel bridges and the remedial measures adopted. These solutions proved successful in the past and the principles may be gainfully utilised for solving future problems as well, if necessary by incorporating modifications to suit the particular details in each case.

7.2 CORROSION AND PITTING

7.2.1 General

One of the most common problems in existing old steel bridges is the damage due to corrosion. The basic cause of corrosion has been briefly discussed in Chapter 2. The extent of damage to steel bridges due to corrosion largely depends on the type of joint details adopted, enveloping environment and quality of maintenance work done.

Some of the locations which suffer from corrosion and pitting are enumerated hereunder.

- In railway bridges, the section of cross girders and stringers which are exposed to waste water from coaches are normally prone to corrosion.
- Surfaces below the sleepers of railway deck do not get satisfactory cleaning and repainting during routine maintenance work. As a result, the steel surfaces in these locations remain wet for a longer period compared to other exposed areas and get pitted due to accelerated corrosion.
- In road bridges, where water or chemicals enter below the deck slab, the steel structure underneath gets damaged due to corrosion.

- Near the end joints of the girders and also at the locations of bearings, dust and debris tend to accumulate. During rainy season these areas remain wet for a longer period causing the steelwork to corrode.
- Bridges located in humid environment, near the sea or in marshy areas and in atmosphere frequently surcharged with smoke, soot or chemical fumes, may suffer accelerated corrosion all over the structure.

7.2.2 Rehabilitation Schemes

The scheme for rehabilitation of a corrosion damaged member depends largely on the degree of corrosion and its extent over the surface area. Where loss of sectional area is appreciable and the capacity of the member is reduced beyond allowable limit, the reduction in sectional area can be made up by fixing additional steel material. Selection of fasteners for fixing the additional material calls for special attention. In riveted bridges, existing rivet holes are frequently utilised for this purpose, by first removing the rivets and then fixing bolts in their place—one at a time. Use of high strength close tolerance bolts are generally preferable over other types of bolts. Since close tolerance bolts can be made to fit tightly (slip free) to match the existing rivet holes, their behaviour pattern becomes somewhat identical to that of the other adjoining existing rivets, thus ensuring satisfactory transmission of forces to the additional material. This as well as other aspects relating to different types of bolted and welded connections have been discussed in Chapter 5. These may be studied before finalising the selection of fasteners.

A typical rehabilitation scheme for a corrosion damaged top flange of a riveted girder is shown in Fig. 7.1. In this case a corrosion plate is provided over the top flange and secured to the top flange by bolts using the holes of the existing rivets. A similar scheme can be adopted for repair of the bottom flange also.

Since web plates of girders and beams are comparatively thin, it is necessary first to accurately establish the extent and degree of the damage. In extreme condition, the web plates may be entirely corroded, resulting in formation of large openings. Generally, web plates damaged due to corrosion can be repaired by fixing corrosion plates of adequate thickness, preferably on both sides of the web secured by bolts. The details and size of the corrosion plates will depend on the location, extent and degree of the damage. Rehabilitation of the corroded web plate should be adequately designed. Figure 7.2 shows some examples of rehabilitation of corroded web plates or girders at different locations. It may be noted that some existing rivet holes have been utilised in addition to drilling new holes for fixing the corrosion plates by means of bolts.

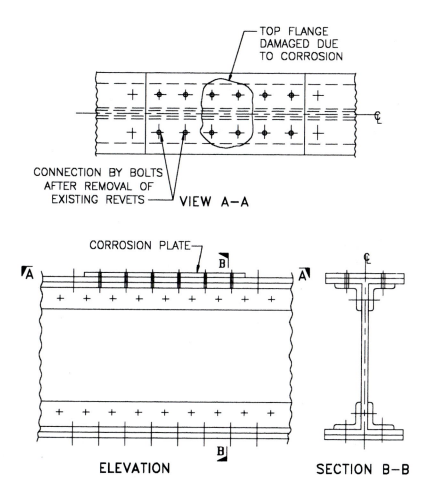

Fig. 7.1: Rehabilitation of corrosion damaged top flange plate of a riveted girder

A typical example of repair of a corroded bottom chord of a truss type bridge has been shown in Fig. 7.3. In some of the older truss type bridges the bottom chords are composed of built up 'U' shaped forms. These chords often get corroded from inside due to accumulation of water. Figure 7.4 shows a method for rehabilitation of the damaged web plates of such a chord.

Damage to secondary members such as gussets, lateral bracings etc., does not directly affect the operation of a bridge. As a result these members are often neglected during routine maintenance work. Corrosion of secondary members may, however, impair the capacity of the main members and also affect the pattern of load path of a bridge. It is, therefore, important to take appropriate countermeasures to rehabilitate secondary members which

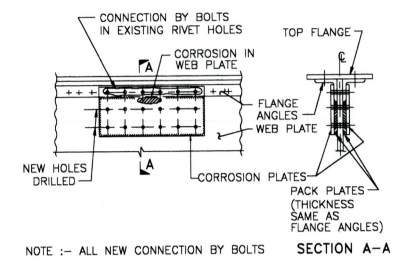

NOTE :— ALL NEW CONNECTION BY BOLTS   **SECTION A—A**

CORROSION NEAR TOP FLANGE

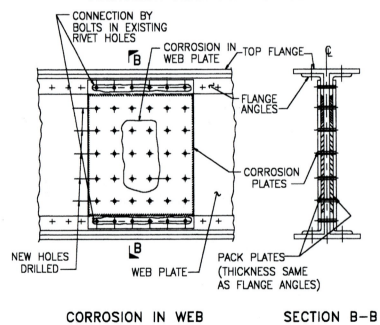

CORROSION IN WEB   SECTION B—B

Fig. 7.2: Rehabilitation of corrosion damaged web plates of riveted girders

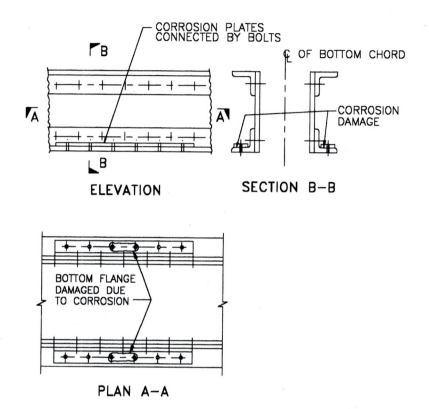

CORROSION PLATES
CONNECTED BY BOLTS

℄ OF BOTTOM CHORD

CORROSION DAMAGE

ELEVATION

SECTION B–B

BOTTOM FLANGE
DAMAGED DUE
TO CORROSION

PLAN A–A

Fig. 7.3: Rehabilitation of corrosion damaged bottom chord of a truss bridge

show signs of distress due to heavy corrosion. It is generally easier to replace or remove for repair such members even during traffic, as they do not affect immediate safety of the structure. Figure 7.5(a) illustrates a typical example of corrosion in gussets and lateral bracings. If the corrosion and pitting in the gussets and connecting rivets are significant, the gusset should be replaced by a new plate of same thickness and connected by bolts. The figure shows that one of the bracing angles is also damaged at the connection due to corrosion. This can be repaired by removing the damaged portion of the angle and replacing the same by an angle of same section and providing splice angles or plates of adequate strength and connecting by bolts as has been shown in Fig. 7.5(b). When a lateral bracing is locally corroded needing rehabilitation, this can be done by fixing two corrosion plates on the bracing angle by means of bolts. However, when a bracing member is badly corroded at a number of locations or where the loss of

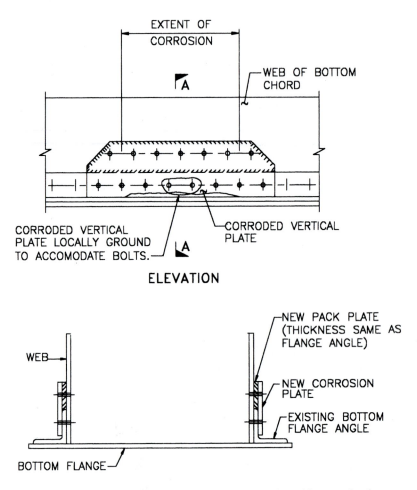

Fig. 7.4: Rehabilitation of corrosion damaged 'U'-shaped bottom chord

sectional area exceeds 30%, it is preferable to replace the member along with the gusset plate.

Generally corrosion in rivets occurs along with the corrosion of members connected by the rivets. These rivets should be replaced by bolts along with the rehabilitation of the member itself. There are, however, cases where only the rivets get damaged badly, although the connecting members remain unaffected. Instances of such damages may be found in the rivets located on the upper flanges of plate girders in deck system of railway bridges, where maintenance underneath the timber sleepers is not easy to perform. Also, in areas with a humid climate, dew formation due to condensation at the underside of riveted components may damage the rivet heads.

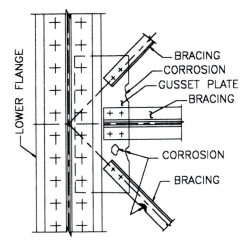

(a) EXISTING DETAIL

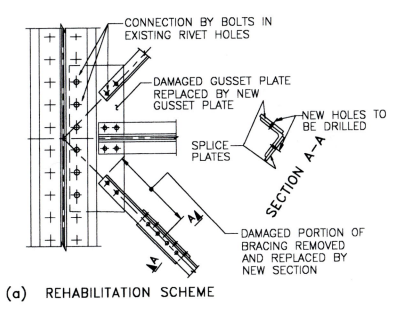

(a) REHABILITATION SCHEME

Fig. 7.5: Rehabilitation of corrosion damaged lateral bracings

Figure 7.6(a) shows a typical rivet head which has been damaged due to corrosion. If the corrosion continues from one side of the connecting plates to the other side, the rivet gets loose and the rivet head on the other side also gets corroded. This has been illustrated in Fig. 7.6(b). In such a

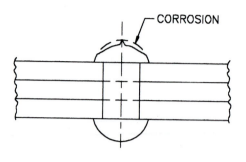

(a)

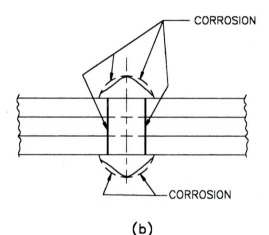

(b)

Fig. 7.6: Corrosion in rivet

condition the rivet should be replaced by a bolt. The normal practice is that if 50% or more of the rivet head gets damaged, the rivet should be replaced by bolt.

7.3 CRACKS

7.3.1 General

Locations damaged by corrosion are normally prone to cracks due to localised overstressing. Cracks also originate in other areas such as connections with inadequate details, roots of flange angles, bent plates etc.

7.3.2 Rehabilitation Schemes

In general cracks occurring in isolated locations can be rehabilitated by any one of the following methods:

(a) First, ascertain the crack length and drill a smooth hole (12–25 mm diameter) at about 20 mm beyond the tip of the crack, along its assumed line of further progress, to arrest further propagation of the crack. Then provide cover plates or cover angles with adequate number of bolts on either side of the crack. This can be done without removing the member and is a very common solution for isolated cases of cracks.

(b) If there are too many cracks in a single member, replace the cracked member entirely by an identical member. Alternatively, a portion of the member which has been damaged may only be replaced. Care should be taken to develop adequate splice connection between the new portion and the existing member.

Cracks resulting from corrosion are generally found at the lower flange of beams of girders near the supports. In riveted bridges such cracks are normally associated with rivets loosened due to corrosion. Figure 7.7 shows an example of cracks developed in the lower flange angles of the main girder of a bridge near the bearing and a suggested retrofit solution by replacing the damaged portion of the flange angles and providing splice angles.

Another example of a crack in a stringer beam near the end support is shown in Fig. 7.8(a). In this case the stringer is located at the bottom flange of the cross girder. The bottom flange of the stringer is badly corroded at this location and a crack has developed from the bottom flange and has propagated diagonally into the web of the stringer end. Suggested retrofit solution is also shown in the figure. In this case, the angle cleats are to be replaced by two larger bent plates to act both as cleats and splices. The connection detail by bolts should be adequate to transmit the forces from the stringer to the cross girder. A new cover plate underneath the bottom flange is also to be fixed by bolts. Figure 7.8(b) shows a retrofit solution for a crack in the web caused due to sharp end of a coping or notch. In this case the existing cleat angles have been replaced by bent plates in the similar manner as has been shown in Fig. 7.8(a). Also a hole at the tip of the crack has been drilled to arrest propagation of the crack.

Welded girders suffer from some typical cracks which are normally not noticed in riveted girders. One such typical crack is the crack developed near the lower end of the welded stiffener connection to the web as shown in Fig. 7.9(a). Normally, these cracks originate at the end of the weld and propagate along the web plate. These cracks may be attributed to residual stress concentration due to welding process at the time of fabrication and subsequent cyclic out-of-plane movement and high bending stresses in the

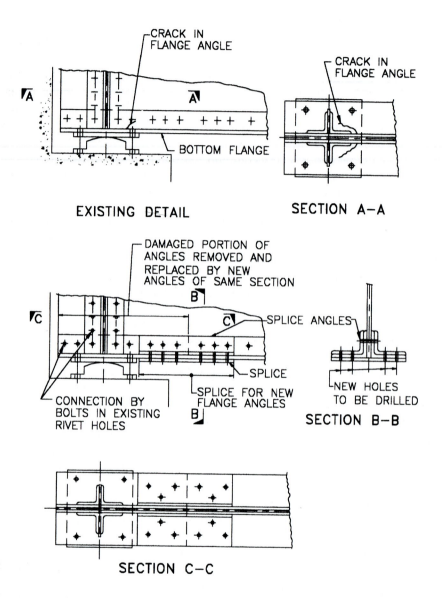

EXISTING DETAIL

SECTION A—A

SECTION B—B

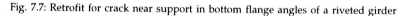

SECTION C—C

Fig. 7.7: Retrofit for crack near support in bottom flange angles of a riveted girder

area of the unstiffened gap in the web leading to fatigue condition and subsequent crack. These cracks can be repaired by first drilling smooth holes about 20 mm beyond the tips of the crack and then gouging out the cracked portion and depositing weld metal in its place. Subsequently the

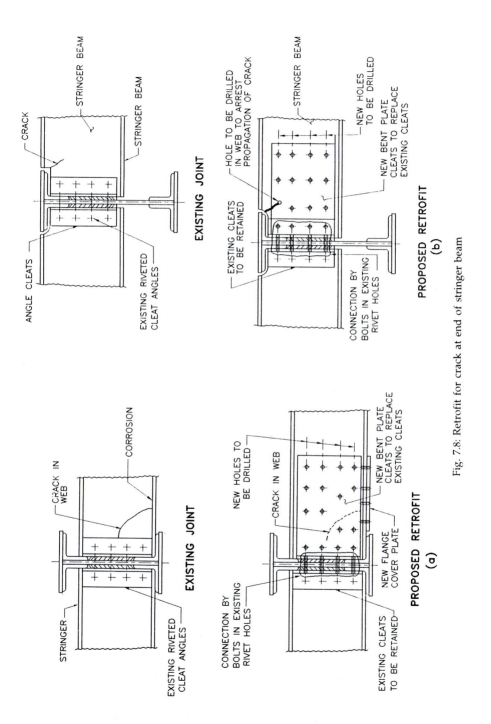

Fig. 7.8: Retrofit for crack at end of stringer beam

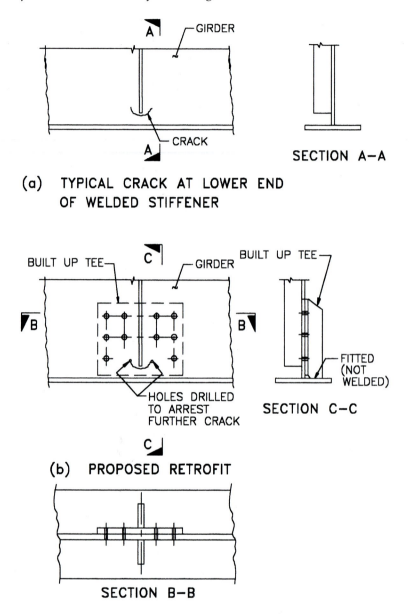

(a) TYPICAL CRACK AT LOWER END OF WELDED STIFFENER

(b) PROPOSED RETROFIT

Fig. 7.9: Rehabilitation of crack in web of welded girder

deposit is finished by removing the excess metal by grinding. Finally, the damaged portion is to be strengthened by fixing built up tee member on the other face of the web plate by bolts as shown in Fig. 7.9(b). The size of the tee member should be sufficient to fully cover the cracked portion.

Another typical example of fatigue cracking is found at the end of an additional plate welded to the bottom flange of rolled or built-up beam. Studies suggest that the crack may be attributed to concentration of stress due to abrupt change in the cross sectional area of the flange. The crack may have originated at the toe of the weld and slowly progressed through the flange. Retrofit solution of such a crack would be to first drill smooth holes at the crack tips to arrest further propagation of the crack. Then the cracked portion should be gouged out and re-welded. The weld should be ground to a smooth profile. Finally, splices similar to the usual beam splice should be fixed by bolts. In case of cracks in the welds connecting a flange to a web, the distress can be repaired by a similar process and provided with splice angles connected to the web and the flange by bolts.

7.4 BUCKLING AND BENDING OF MEMBERS

7.4.1 General

Buckling and bending of members due to collision or accident are very common. If the bending is local and small, it may be straightened either by mechanical straightening or by flame straightening.

(a) Mechanical Straightening

There are two methods for mechanical straightening viz. straightening at the ambient temperature and straightening while the damaged area is heated. Straightening in the ambient temperature is generally not recommended, as heavy external loads are required to be applied during the straightening process. This might affect the properties of steel adversely. The recommended process is to heat the area slowly and then straighten it by mechanical means avoiding impact load. It should then be allowed to cool without any external aid.

(b) Flame Straightening

The methodology for fabrication of bent beams can be conveniently used for straightening bent members also. In this method the principle of thermal expansion and contraction is utilised for straightening a bent member by applying concentrated heat from a blow pipe to a wedge shaped area on the member. When a bent member is heated uniformly at a particular location, the heated portion will tend to expand and this expansion will be prevented by the surrounding cooler metal. The forces hindering the expansion will cause the relatively weak heated part to expand (bulge) externally. When the area cools, the contraction causes the bend to straighten. The temperature should be restricted to approximately 700°C i.e., dull red.

The heating should proceed from the base to the apex of the wedge and the heat should penetrate evenly through the plate thickness, maintaining an even temperature. Figure 7.10 illustrates the methodology of flame straightening of bent members. Although the method appears to be quite simple, it needs sufficient field experience and practice to be successful. As a result, the method though very effective has not gained popularity amongst technicians.

7.4.2 Rehabilitation Schemes

There have been many instances of bottom chord of a trussed girder or bottom flange of plate girder road over bridges having been damaged from vehicular collision due to low clearance of the structure without adequate protection or due to driver error.

Figure 7.11 shows an example of such a damage in a riveted plate girder bridge. In this case the bottom flange plate has been damaged at a number of locations. Three alternative schemes can be considered for rehabilitation of the bridge:

(a) The entire bottom flange can be removed and replaced by a new flange plate of same quality and section and fixed to the bottom flange angles by bolts.

(b) Instead of removing the entire flange plate, only the damaged portions of the bottom flange plate can be removed and replaced by new plates of requisite lengths at these locations and fixed by bolts. Adequate splice plates should also be provided for transmitting forces to the existing flange.

(c) If the distortion is not significant, the damaged portions of the bottom flange plate can be straightened and additional flange plate fixed below these portions by bolts. These plates should extend sufficiently beyond the damaged portions of the flange plate, so as to accommodate re-quired number of bolts for transmitting the load to the undamaged bottom flange plate.

In all the above schemes the new bolts should match the existing rivet holes. Figure .7.11 shows typical details of solution for schemes (b) and (c) above.

There have also been instances of main members of through and semi-through railway truss bridges suffering extensive damage due to derailment of railway wagons. Somewhat similar situation may also occur due to ex-plosion resulting from war or terrorist activities. If the damages are exten-sive and are likely to impair the bridge, the operation of the line has to be stopped immediately, keeping in mind the safety aspect of the passengers. However, it becomes almost mandatory to open the line again at the earliest possible time, albeit temporarily, and with speed restrictions, so that the

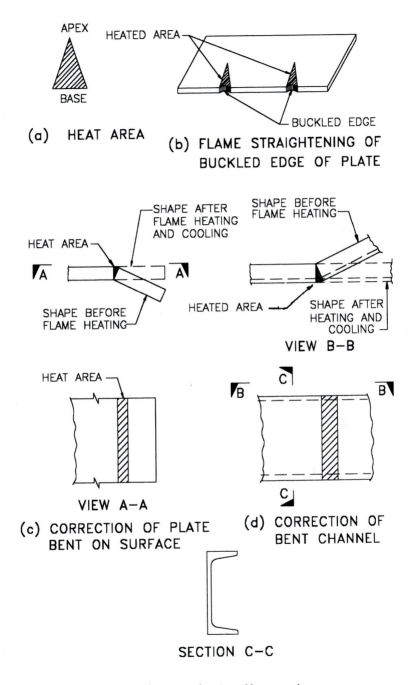

APEX

BASE

(a) HEAT AREA

HEATED AREA

BUCKLED EDGE

(b) FLAME STRAIGHTENING OF BUCKLED EDGE OF PLATE

SHAPE AFTER FLAME HEATING AND COOLING

SHAPE BEFORE FLAME HEATING

HEAT AREA

SHAPE BEFORE FLAME HEATING

HEATED AREA

SHAPE AFTER HEATING AND COOLING

VIEW B-B

HEAT AREA

VIEW A-A

(c) CORRECTION OF PLATE BENT ON SURFACE

(d) CORRECTION OF BENT CHANNEL

SECTION C-C

Fig. 7.10: Flame straightening of bent members

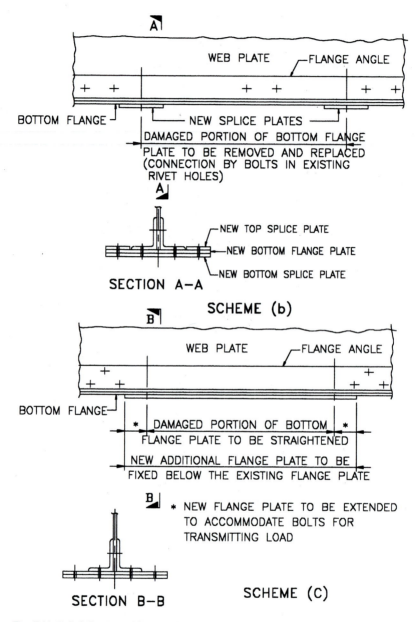

Fig. 7.11: Rehabilitation of bottom flange of girder damaged by vehicular collision

social, commercial, economic or defence needs are met to some extent. In such cases, the rehabilitation has to be planned in two stages, first, a temporary restoration of traffic followed by a permanent rehabilitation. Figure 7.12

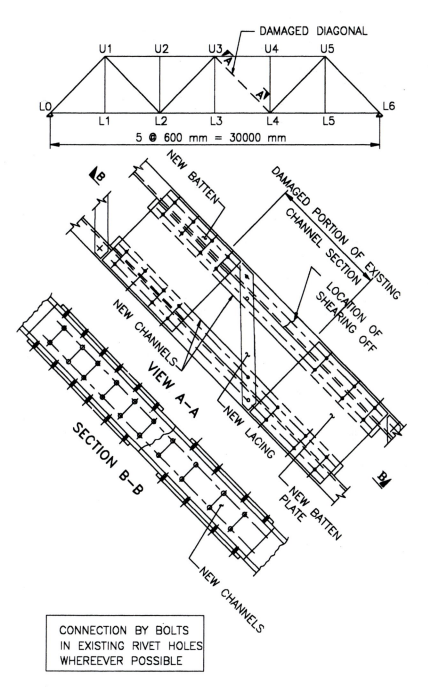

Fig. 7.12: Restoration of damaged diagonal

illustrates an example of such a damage to one truss of a riveted railway bridge and the counter-measures for rehabilitation of the bridge. Inspection and survey work have to be carried out almost immediately after the accident, to ascertain the exact nature and extent of the damage. In this case, diagonal U3-L4 made of two channel sections has suffered maximum damage viz., the trackside channel section got sheared off because of excessive twist. The lacings in this location have also been twisted or torn. The adjacent verticals and diagonals, and also top flanges of some cross girders suffered minor bends or twists at different locations.

In all such cases, it is important to ascertain which of the members or parts of members have been overstressed beyond yield point and deformed permanently. Repairs should be carried out only on members where average stress level has remained below yield stress.

Temporary counter-measure: First the damaged lacings of the diagonal U3-L4 are to be removed and the twisted channel section is to be straightened. Two channels of identical section are then fixed on the backs of the existing channel, the length extending beyond the damaged portion, so that adequate number of bolts to transmit the forces can be accommodated. New battens/lacings are to be fixed to the member as shown in the figure. The localised damages to adjacent vertical diagonals and cross girders should be rectified by local strengthening.

Permanent rehabilitation: Permanent deformation in the channel section of the diagonal conveys that the steel material has gone beyond the yield point and as such its contribution to elastic behaviour of the structure is uncertain. This condition also makes the member more vulnerable to fatigue failure. Therefore, as a permanent rehabilitation measure, the deformed member should be removed and replaced by a new one at the earliest opportunity. As mentioned earlier, the different members of the bridge structure should be relieved of the dead load stresses. The most common method for this is to provide temporary supports at the different node points of the bottom chord. Once the bridge structure is made to rest on these node points, the concerned member may be removed and a new pre-fabricated member erected in its place.

7.5 INADEQUATE END CONNECTIONS

7.5.1 General

Distress due to inadequate end connections is noticed mostly in the deck system of bridges.

7.5.2 Rehabilitation Schemes

Normally, stringers in bridges are designed as simply supported beams. However, in many cases, the end connections are detailed in such a way that an unintended continuity is introduced at these locations, which is not desirable. Figure 7.13(a) illustrates such a situation, where the two stringers in a railway bridge rest on two seating brackets fixed on a cross girder and are connected to the cross girder by double cleat angles by rivets. These rivets are designed to transmit only the end shears to the cross girder. However, since these rivets connect the two stringer ends through the web of the cross girder, continuity is developed between these two beams. Consequently the connection becomes inadequate to transmit the moment due to the unintended behaviour of the beam ends and the rivets become loose. The joint can be strengthened by replacing the two seating brackets fixed on the cross girders by comparatively more rigid built-up brackets and connecting these to the stringers by adequate number of bolts. The recommended scheme is shown in Fig. 7.13(b).

In some old bridges the stringers are placed over the cross girders. Here, again, the joints at cross girders should be adequate to cater for the support moments of the continuous stringers. Figure 7.14 shows an example of distress in the connecting rivets at such a location and a suggested scheme to improve the behaviour of the connection. In the scheme, stiffeners have been introduced on the cross girders as well as on the stringers. Also stiffening members have been added at these locations to improve the rigidity of the joint. In view of the possibility of fatigue condition of this area due to the change of stress pattern, welding should be avoided. Connection with bolts in therefore recommended.

7.6 UPGRADING OF EXISTING BRIDGES

7.6.1 General

When the loading standards of a country or a region are modified due to introduction of heavier vehicles or railway locomotives, it becomes necessary to check the capacity of the existing bridges in the country or region for the increased loadings and upgrade (i.e., increase live load carrying capacity) these bridges if necessary. It is quite possible that many of the existing bridges which were designed earlier by conservative methods (such as, uni-plane truss with pin connected members) may be found to be structurally adequate when modern methods of space frame analysis with computers and present day design standards are applied for checking. In some cases the checking may reveal that only a few of the members are inadequate and need strengthening, while the rest of the members and connections are in order.

In general, the procedures available for upgrading are:

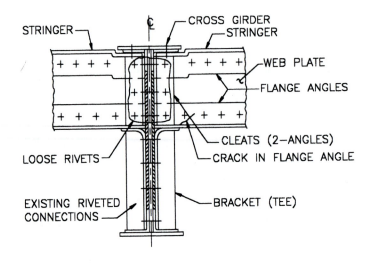

(a) EXISTING JOINT

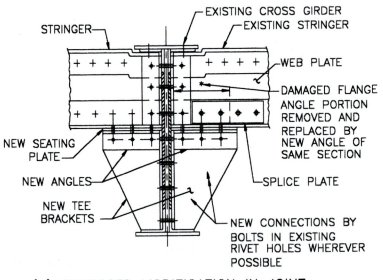

(b) PROPOSED MODIFICATION IN JOINT

Fig. 7.13: Rehabilitation of end connection of stringer beam

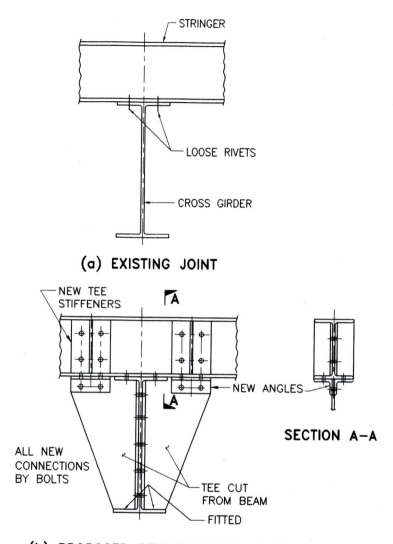

Fig. 7.14: Improvement of end connection

(a) Strengthening of the deficient members
(b) Introduction of supplementary members
(c) Reduction of dead load
(d) Modification of existing structural system
(e) Replacement of the deficient members

7.6.2 Strengthening of Deficient Members

One of the most common methods of strengthening beams and girders is by adding cover plates to the top and bottom flanges near the midspan. The length of the cover plates and their cut-off points are to be determined by design check. In case of rivetted girders, the cover plates are to be pre-drilled to a smaller diameter, reamed *in situ* after stitching and bolted at site by close tolerance bolts. The holes in the cover plates must match the disposition and diameter of the existing holes. Similarly, permanent bolts should also be inserted by removing the temporary pins, one at a time. This procedure would ensure smooth replacement of rivets by bolts without any mismatch in the holes.

In case of rolled beams or welded girders where the cover plates can be welded, it is preferable to extend the plates to the full length of the beam or the girder instead of terminating these at the theoretical cut-off point. This arrangement would minimise the chance of fatigue related cracks due to sudden change of stress at the ends of the cover plates, resulting from change in the cross sectional area at these high-stress locations. Also, the width of the top flange cover plate should be narrower and that of the bottom cover plate wider than the corresponding flanges, in order to ensure down-hand welding at site. Figure 7.15(a) illustrates the arrangement.

For a girder which supports concrete deck on the top flange, welding an additional plate to the top flange calls for removal of the reinforced concrete slab over the width of the flange for the total length of the girder. This work entails elaborate arrangements for providing support to the entire remaining portion of the slab. Also, care has to be taken not to damage the reinforcements while removing the concrete slab over the beam. Obviously this work requires close supervision at site. Therefore, unless the entire deck concrete is required to be replaced, this method is not likely to be cost efficient. An alternative method would be to fix new plates or angles below the top flange as shown in Fig. 7.15(b). This figure also shows some other alternative schemes for increasing the areas of top and bottom flanges. Some of the details shown in this figure may be adapted for riveted girders also.

In many old bridges the bottom chord (tension member) comprises of two main components—mostly built-up channels—but with no lacing or batten between them. Consequently, these components act as two separate tension members without much flexural strength. Providing new lacing system to these bottom chords would substantially increase their flexural capacity by adding to stiffness and would reduce chances of vibration. This scheme has been successfully adopted in a number of railway bridges and the results have been very encouraging. Figure 7.16 shows typical details of such a strengthening scheme.

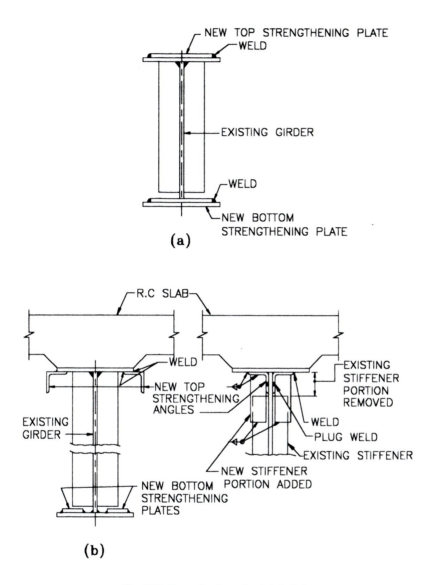

Fig. 7.15: Strengthening of welded girders

In most of the main members such as top and bottom chord, end rakers, diagonals and verticals, if additional area is required to be provided, this can be done by bolting or welding additional plates in the webs or flanges of the members. Some typical details have been shown in Fig. 7.17. Care should, however, be taken to ensure that the centre of gravity of the strengthened section coincides with the original section to avoid fresh secondary stresses due to eccentricity. Figure 7.17(d) shows a detail where the capacity of a built-up top chord has been enhanced by providing ad-

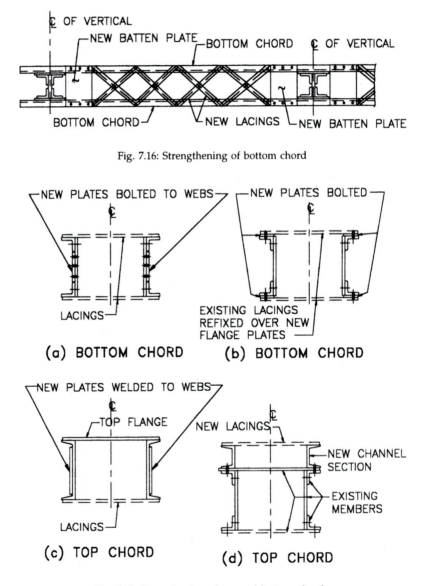

Fig. 7.16: Strengthening of bottom chord

(a) BOTTOM CHORD

(b) BOTTOM CHORD

(c) TOP CHORD

(d) TOP CHORD

Fig. 7.17: Strengthening of top and bottom chords

ditional built-up members over the top chord. Since such additions are likely to shift the centre of gravity line of the chord, effects of secondary stresses due to eccentricity should be considered while analysing the load effects on the truss.

In some older bridges the diagonals which take only tensions, are made up of two individual flat sections only, without any lacing or batten between them. Such diagonals are prone to unwelcome vibrations. Providing lacing system to such diagonals will enhance their rigidity and would improve the behaviour significantly. A typical detail is shown in Fig. 7.18.

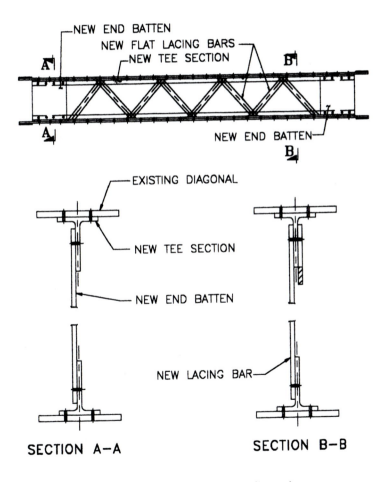

Fig. 7.18: New lacing system in diagonal

7.6.3 Introduction of Supplementary Members

Capacity of top chords or any other compression members of a truss bridge can be increased by reducing the effective lengths of these members. The effective lengths in the plane of the truss can be conveniently reduced by subdividing the panels by introducing new members. This has been illustrated in Fig. 7.19. When effective length of top chord across the plane of the truss is required to be reduced, introduction or modification of top lateral bracing system can provide the solution. However, in a semi-through truss bridge where top lateral bracings cannot be introduced due to insufficient headroom, the effective length of the top chords may be reduced by introducing 'U' frame action through the existing verticals acting with the cross girders at one or more intermediate locations. For this purpose the verticals and their connection details with the cross girders are to be checked and strengthened, if necessary, to provide adequate lateral supports to the top chords at intermediate locations. Figure 7.20 illustrates a typical proposal for strengthening the connection between existing vertical and cross girder.

Figure 7.21 shows how a new 'U' frame system can be introduced in an existing double-warren type semi-through bridge girder.

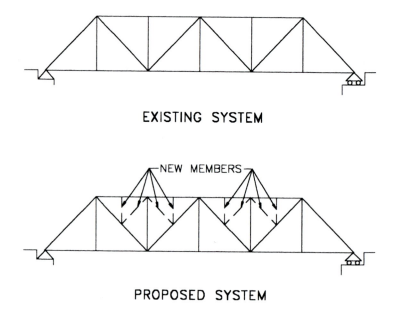

EXISTING SYSTEM

PROPOSED SYSTEM

Fig. 7.19: Strengthening of truss by introduction of new members

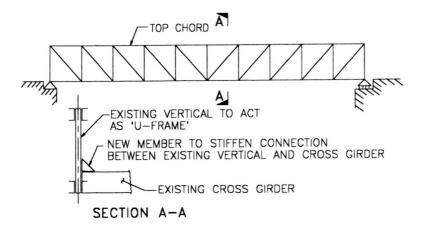

Fig. 7.20: Upgrading by strengthening existing verticals to act as 'U Frames'

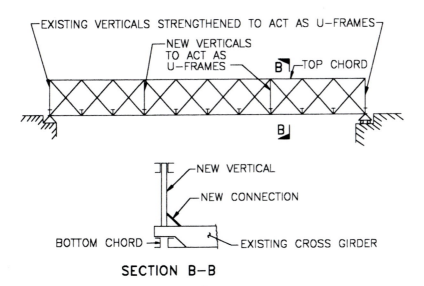

Fig. 7.21: Upgrading by introduction of verticals to act as 'U Frames'

7.6.4 Reduction of Dead Load

Live load carrying capacity of a bridge can be increased by reducing the dead load on the structure. As for example, in a railway bridge with ballasted deck system, where the weight of ballast contributes substantially to the overall load on the bridge, significant reduction of dead load can be achieved by introducting stringers and cross-girders in place of 'ballast-on-trough' deck system. Figure 7.22 shows a typical scheme. In this case the existing system consists of the railway track resting on ballasts over steel troughs which span transversely and are supported on the bottom flanges

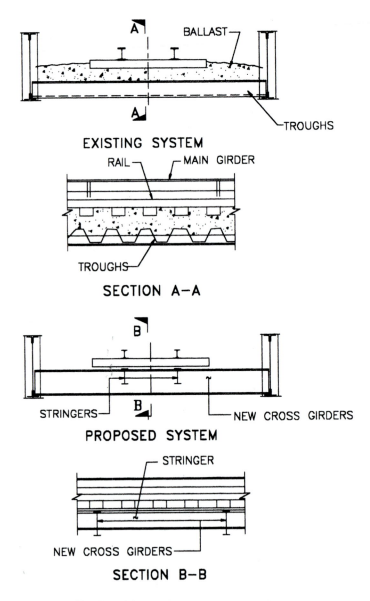

Fig. 7.22: Schemes for reduction of dead load

of main girders. In the proposed alternative scheme, the railway track is supported on longitudinal stringers which span between cross-girders connected to the main girders.

Another example of reduction of dead load in a road bridge with concrete deck supported on steel beam system is by replacing the RC deck by orthotropic steel deck system. Essentially, this system consists of steel deck plate stiffened by steel ribs welded to the bottom of the deck plate. The ribs may be open type or closed type as shown in Fig. 7.23(a). The units can be fabricated to the required dimensions—usually modular design—with overall size to suit the allowable time for removal of existing concrete deck panel and erection of the new steel deck module. The wearing surface of bituminous asphalt pavement (or costlier but more durable epoxy asphalt system) can also be installed in the shop to enable these units to be usable immediately after erection. Figure 7.23(b) shows a typical scheme for replacement of RC deck by orthotropic steel deck in a road bridge. Orthotropic steel modules have been found to be very useful for bridge deck replacement in a number of bridges recently, such as George Washington Bridge over the Hudson River in New York (1978), Golden Gate Bridge in San Francisco (1985), Throgs Neck Bridge in New York (1986), Benjamin Franklin Bridge across the Delaware River in Philadelphia (1987).

The main advantages of orthotropic bridge decks may be summarised as follows:

(a) Reduced Dead Load
The dead load of the deck can be considerably reduced. Consequently, the live load capacity of the bridge can be enhanced with this method. There would be other structural advantages due to reduction in dead load, such as general reduction of stresses in all elements of the bridge including the foundations. Also the seismic resistances will increase due to reduced dead load.

(b) Simple and Easy Erection Work
One of the significant advantages of orthotropic steel plate system is that the individual units or modules are easy and simple to install and match with the existing structures.

(c) Faster Site Work
Since the orthotropic steel modules including the deck can be wholly prefabricated in the workshops, the time cycle between removal of concrete deck in a panel and installation of the steel deck system to be ready for use can be restricted considerably. This makes the orthotropic steel plate system eminently attractive due to ease of rehabilitation of decks in busy and strategic bridges where all traffic lanes must be available during rush

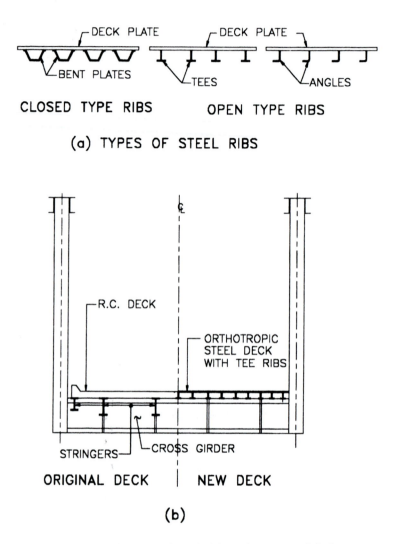

Fig. 7.23: Replacement of RC deck by orthotropic steel deck

hours and repair work is allowed to be carried out only during night time or during a very restricted period of time.

7.6.5 Modification of Structural System

The load carrying capacity of a bridge can be increased by incorporating changes in the details, so as to modify the basic structural system of the bridges. There may be many examples of this method of capacity enhancement. Some of these are described below.

(a) Conversion of Simply Supported Spans to Continuous Span

If in an existing bridge the longitudinal stringers are designed as simply supported and detailed as such, it may be possible to improve the load carrying capacity of these stringers by incorporating modifications in the end connection details. A typical example of the details of such a modification is shown in Fig. 7.24. In this case the existing stringers of a railway bridge are simply supported at the two ends between cross girders. The end shear is transmitted by angle cleats and support brackets fixed by rivets to the web of the cross girders. The modification to increase the capacity of the stringers involves replacing the two seating brackets by comparatively more rigid built-up brackets and providing tension plate over the top flange of the cross girder and connecting this plate to the top flanges of the stringers. The connections are to be done by bolts.

(b) Conversion from Non-composite to Composite Girder System

Carrying capacity of a non-composite girder can be increased by installing new shear connectors on the top flange to enable the girder to act compositely with the deck slab. This system is suitable when the deck slab shows significant distress warranting replacement. After the damaged deck slab is removed, shear connectors may be welded on the top flange of the existing girder and new concrete deck cast.

In cases where the deck slab is in good condition, the concrete can be removed locally and shear connectors installed over the top flange. The local void in the concrete should be filled with non-shrink grout so that the girder can behave as a composite girder. Figure 7.25 illustrates the method.

(c) Providing Additional Support

The capacity of an existing bridge can be considerably increased by providing additional supports from below the bottom chord at one or two points. Figure 7.26 shows a typical example of such an arrangement. Since this solution would require a new support arrangement to be constructed below the existing structure, the reduction in the width of the roadway beneath and also the difficulties likely to be faced at site for such construction are to be considered, prior to finalising the scheme.

Also, shifting of support points from the ends of the bridge to the next inward panel point may increase the capacity of a girder. The new configuration would be a reduced span with cantilevered panel at ends. A schematic arrangement has been shown in Fig. 7.27.

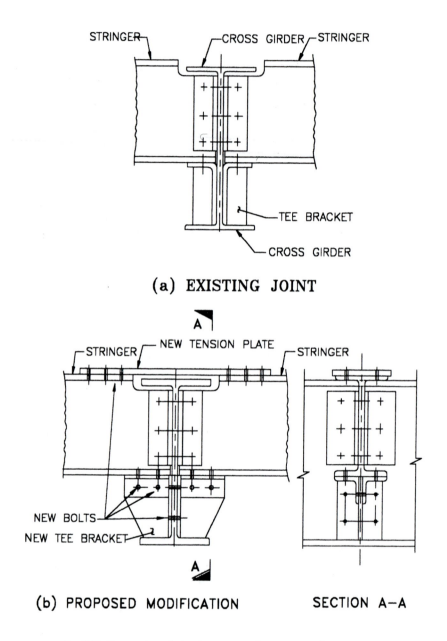

Fig. 7.24: Conversion of simply supported span to continuous span

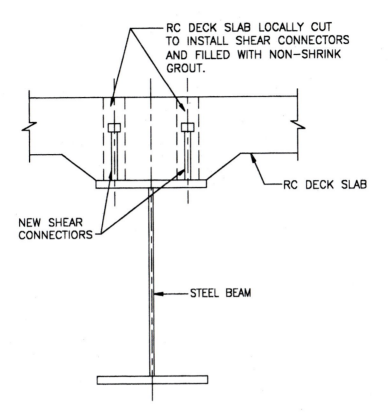

RC DECK SLAB LOCALLY CUT
TO INSTALL SHEAR CONNECTORS
AND FILLED WITH NON−SHRINK
GROUT.

RC DECK SLAB

NEW SHEAR
CONNECTIORS

STEEL BEAM

Fig. 7.25: Conversion from non-composite to composite girder system

(d) Prestressing

The load carrying capacity of an existing bridge can be increased by introducing counterbalancing forces in the system by means of post-tensioning
tendons, which work much the same way as in a post-tensioned concrete
beam. This procedure induces stresses on the structure and reduces the
effects of the existing dead or live load on the structure, thereby increasing
the live load capacity of the bridge. The tendons can be straight or curved
in configuration and can be located either within the structure or outside
it. Figure 7.28(a) shows a common method of prestressing a beam or a
girder by placing two straight tendons or H.T. bars on two sides of the
web a little above the bottom flange and anchoring these tendons or bars
at the ends. The tendons or bars can also be placed below the bottom flange
and anchored to it at the ends as shown in Fig. 7.28(b). An illustration of
tendons located on two sides of the web of a girder and draped below the

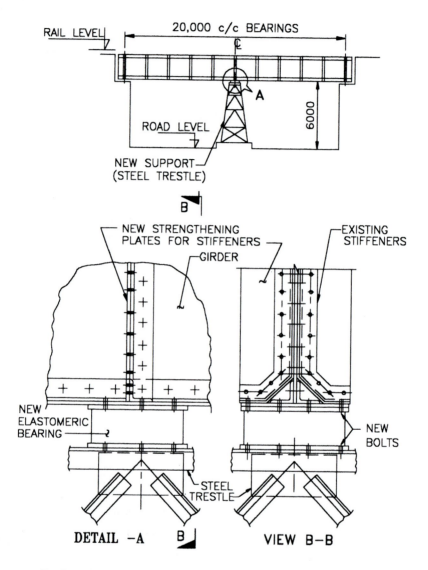

Fig. 7.26: Capacity increase by providing additional support to girder

centre of gravity of the girder cross-section at midspan is shown in Fig. 7.28(c). In this case the tendons are made to pass over saddles fixed on the web of the girder to provide the required curvilinear configuration. The ends of the tendons are fixed to anchor headstocks located above the centre of gravity. Figure 7.28(d) shows another detail in which a king post saddle

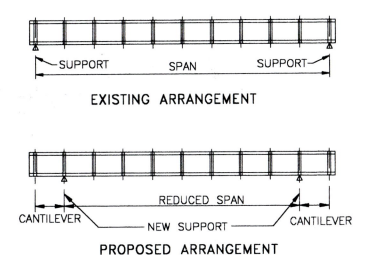

EXISTING ARRANGEMENT

PROPOSED ARRANGEMENT

Fig. 7.27: Capacity increase by shifting end supports

support is fixed below the bottom flange of the girder at midspan and the tendon is passed over the saddle and connected to anchor head stocks located at the end of the girder.

In the above methods, the eccentricity of the tendon induces a hogging moment on the girder which counterbalances the sagging moment caused by the dead and live loads. A typical stress distribution diagram in a beam at the cross-section of maximum bending moment, when post tensioning is applied, is shown in Fig. 7.29. The beam is already loaded on the top flange by a uniformly distributed load, which produces a flexural stress of f_0 at the top and bottom fibres of the beam. An eccentric horizontal force from tensioning of the tendons produces normal compressive stress f_1 and bending stress f_2 at the extreme fibres of the beam. It will be noted from the diagram that the bending stress due to tensioning of the tendons relieves the bending stress due to service loads in both the tension and compression flanges. Thus the beam after post tensioning is capable of taking more live load.

As prestressing method induces high compressive force, it is important to check the ability of the existing member to withstand the effect of such force without local or overall buckling. The present condition of the concerned member thus becomes an important aspect to examine. Strengthening of the member is to be done if necessary. Also the effect of prestressing on the other members of a bridge girder must be considered in this respect.

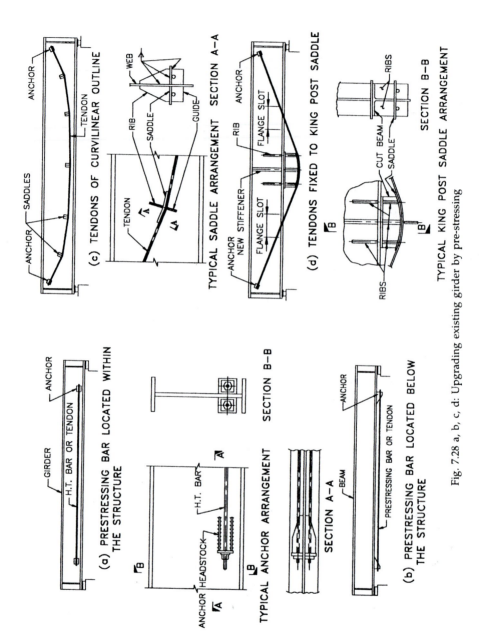

Fig. 7.28 a, b, c, d: Upgrading existing girder by pre-stressing

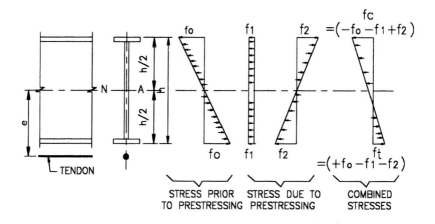

Fig. 7.29: Stress distribution due to prestressing

7.7 INADEQUATE CLEARANCE

Existing through type bridge girders may need modifications to accommodate increased vehicle clearance requirements due to various reasons such as introduction of new types of locomotives, container services, electrified traction etc. End portal systems and sway bracings are the most common members that are affected due to such requirements. A few suggested modifications in the end portal system have been shown in Fig. 7.30. There may be many other solutions also. In all such cases, design check should be carried out to ensure that the system adopted is adequate to transmit the lateral forces from the top chords to the bearings.

7.8 WIDENING

Widening of existing bridge structure is generally associated with widening of the connecting road network to accommodate increased traffic and consequent additional traffic lanes. Sometimes widening is also done to satisfy specific modifications in the traffic clearance dimensions of the concerned authorities. There are a few traditional methods for widening of a bridge. These will be briefly discussed in the following paragraphs. In all cases, however, great care should be taken in the design of the widened structure, as the stiffness and load distribution characteristics of the structure are likely to change. Special attention should be given to the connections—existing as well as new—for providing compatible load sharing arrangement due to changed configuration of the structure.

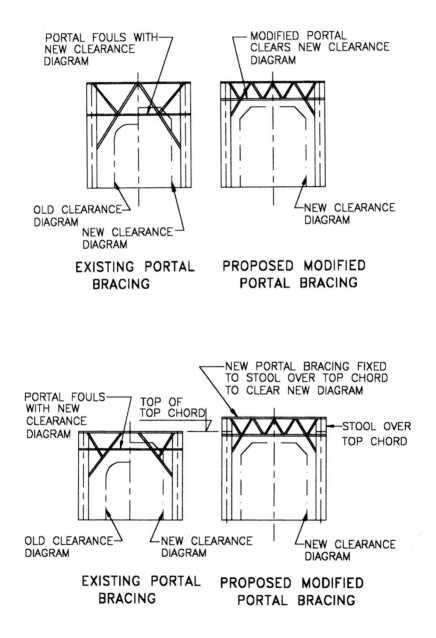

Fig. 7.30: Modifications to existing portal bracings

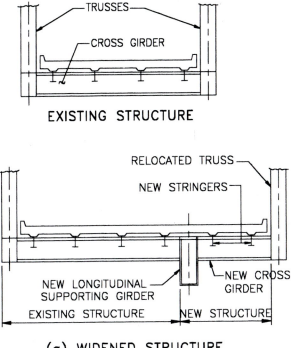

EXISTING STRUCTURE

(a) WIDENED STRUCTURE

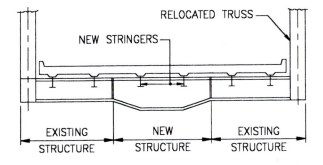

(b) ALTERNATIVE ARRANGEMENT

Fig. 7.31: Schemes for widening of bridge

One of the common methods of widening an existing bridge is to relocate one of the trusses further away from the other truss and introduce a new longitudinal supporting girder near the previous axis of the relocated truss. Figure 7.31(a) illustrates the arrangement. The existing floor system thus spans between the new longitudinal girder and one of the existing trusses. The relocated truss and the new longitudinal girder in turn support the added width of the deck. With this arrangement, the new longitudinal girder has to be designed for its share of the load from both the existing as well as the added width of the deck.

An alternative arrangement has been shown in Fig. 7.31(b). Here, instead of adding a new longitudinal supporting girder, the cross girders of the bridge are lengthened by cutting these in the middle and inserting additional lengths between the two halves of the existing ones. Longitudinal stringer beams are to be added in the middle portion to support the added deck. In this arrangement the modified cross girders are to be checked for the increased span. Also, the main trusses may need strengthening for the increased dead and superimposed loads.

Figure 7.32 shows another arrangement in a deck type bridge where the two existing girders have been moved closer together and two new girders added on the outer sides. With this arrangement, the existing girders may

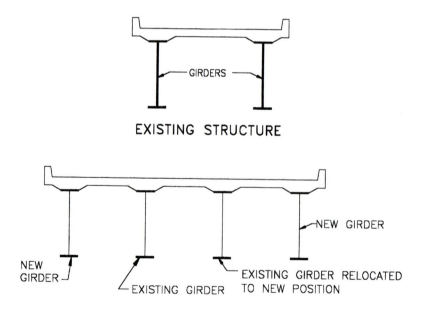

Fig. 7.32: Widening of bridge by adding longitudinal girders

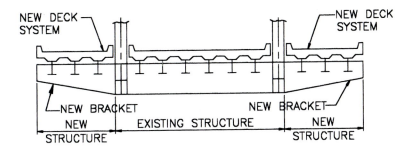

Fig. 7.33: Widening of bridge by addition cantilevered brackets

not need any strengthening. In some cases, the existing girders may have enough strength to support increased deck width even without needing relocating or strengthening.

Additional lane may also be provided by constructing suitable cantilevered brackets on the outsides of the trusses (Fig. 7.33). In this arrangement the trusses may need strengthening due to the increased loads. Use of light weight orthotropic steel deck system may reduce the dead load of the deck necessitating only nominal strengthening or eliminating strengthening altogether for the trusses.

Construction of an independent structure by the side of the existing bridge may perhaps be the easiest solution. However, cost aspect of such a solution needs careful study.

Bibliography

1. Swindlehurst, J and Parkinson, FH: 'Steel Structures', *Bridge Inspection and Rehabilitation*, Parsons Brinkerhoff, John Wiley & Sons Inc., USA, 1993.
2. Park, Sung H: *Bridge Rehabilitation and Replacement*, S.H. Park, Trenton, New Jersey, USA.
3. *Handbook of Maintenance, Inspection and Rehabilitation of Railway Structures*, United Nations Economic and Social Commission for Asia and the Pacific, 1990.
4. Stahl, FL: 'Orthotropic Steel Plates for Bridge Deck Replacement', *Extending the Life of Bridges*, "ASTM:STP 1100, GW Maupin, Jr., BC Brown and AC Lichtenstein, Eds., American Society for Testing and Materials, Philadelphia, 1990, pp 109–120.
5. *Control of Welding Distortion*: The Institute of Welding, London, 1957.
6. Belenya, E: *Prestressed Load Bearing Metal Structures*, MIR Publishers, Moscow, 1977.

8

Case Studies

8.1 INTRODUCTION

In this chapter case studies are presented to illustrate some interesting techniques which were developed or innovated to overcome problems encountered in a number of steel bridges. The details of these techniques have been collected from various published literature as well as from the author's personal notes on the subject. References of the source of the information presented have been given at the end of each case study.

8.2 RAILWAY BRIDGE OVER SUNGAI KERAYUNG, MALAYSIA

Located at Kuala Lumpur, Malaysia, this double metregauge track Pratt truss bridge is of about 31.3 m span. The bridge has some very unusual details compared to modern bridges. The deck system is below the level of the bottom chords—the stringers carrying the railway tracks are connected to the cross girders, which are hung at the ends from the bottom chords of trusses. Before the rehabilitation work in 1995, the top lateral bracing system consisted of cross beams fixed over the top chords with diagonal bracings between the top flanges of some of these cross beams (in four bays only). These cross beams also extended about 1.2 m beyond the top chords and were connected to the verticals from outside by knee bracings. There were no portal bracings at raker ends. Also, some of the diagonals and bottom chords consisted of two separate sections without any lacings.

Prior to rehabilitation this bridge was notorious for its excessive oscillation during passage of trains. Since Malaysian Railways were due to introduce new locomotives with 20T axle loads to run at a speed of 120 kmph, the bridge was checked for its capacity to bear this kind of static and dynamic loads. The checking revealed that the arrangement of the existing top lateral bracings (particularly the absence of end portals) was inadequate

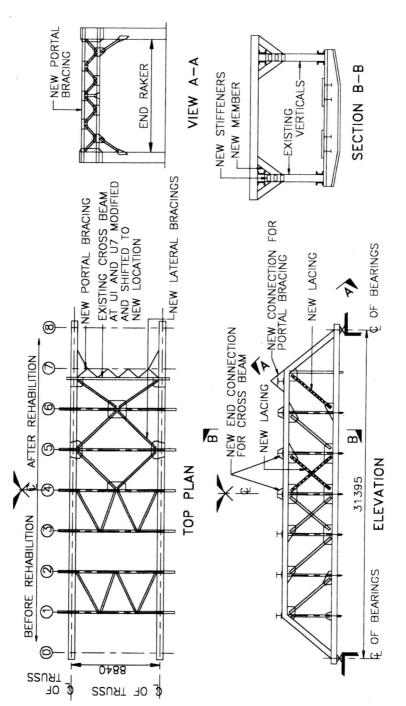

Fig. 8.1: Rehabilitation of railway bridge over Sungai Kerayung, Malaysia

to transmit the lateral forces to the bearings. Also, absence of lacings or battens in some diagonals and bottom chords contributed to the distress.

The bridge was rehabilitated by introducing new top lateral bracing system between the existing cross beams. To improve the horizontal rigidity of the bridge, the connections of the existing cross beams with the top chords were strengthened. Internal sway bracings were also added. As regards portal bracings, the traditional system of portal knee bracings on the end rakers were not feasible, as these interfered with the minimum structure clearance gauge for future locomotives and the proposed electrification of tracks. In view of these restrictions the portal bracings at raker ends were placed above the top chords and fixed on raised support stools with identical slope as the rakers.

Also the bottom chords in four panels did not have any lacing or batten between the two main components of the built up channels. The flexural capacities of these chords were found to be inadequate. New lacings were added to these chords to increase their capacities.

Similarly, in six panels the diagonals were made up of two individual flat sections only without any lacing or batten between them. Lacings were added to these diagonals in order to enhance their rigidity and thereby improve their behaviour.

Figure 8.1 shows the rehabilitation scheme for the bridge.

(Source: Ghosh, UK, Das, PP and Ghoshal, A: 'Rehabilitation of Steel Bridges—Experiences in India and Malaysia', *The Bridge & Structural Engineer*, Indian National Group of the International Association for Bridge & Structural Engineering, New Delhi, March 1997.)

8.3 RAILWAY BRIDGE AT SALAK SOUTH, KUALA LUMPUR, MALAYSIA

This railway bridge is located in Kuala Lumpur and is carrying two ballasted railway tracks over a busy roadway underneath. It is a single span semi-through plate girder bridge of about 11 m overall length. The superstructure essentially consists of three longitudinal plate girders between the two abutments with transverse trough sections placed on the bottom flanges of these longitudinal girders. The clear height from the roadway is approximately 4.0 m.

The roadway underneath the bridge has two carriageways (up and down traffic) serving both light and heavy vehicles. Located at a strategic position, the road is very busy and full of traffic at peak periods. Due to inadequate headroom and in the absence of any device to restrict vehicles with height beyond the allowable limit, the bottom flange plates of the two outer girders were badly distorted due to vehicular collisions at locations of incoming

(a)

(b)

Fig. 8.2: Road over bridge at Salak South, Kuala Lumpur, Malaysia: (a) typical view of the
damages, (b) bridge after rehabilitation

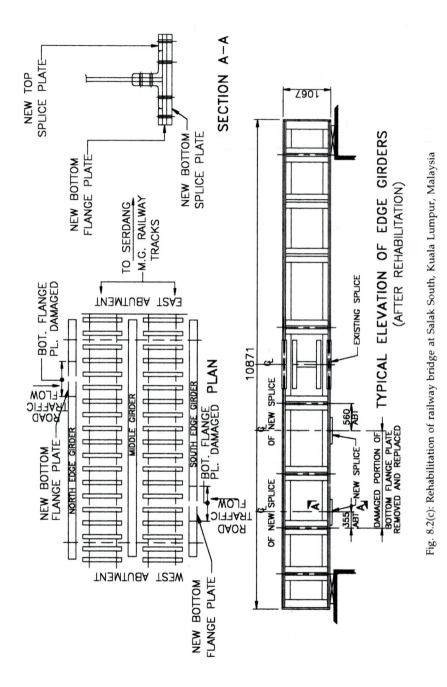

Fig. 8.2(c): Rehabilitation of railway bridge at Salak South, Kuala Lumpur, Malaysia

traffic flow (left hand sides). Figure 8.2(a) shows a typical view of the nature of the damages.

Since the roadway below is a major link in the existing traffic flow system in the area, it was not possible to close the roadway to traffic for carrying out rehabilitation work. Keeping this in view, two alternative solutions for rehabilitation of the damaged outer girders were considered:

(a) The distorted portions of the bottom flange plates to be straightened and additional strengthening plates to be fixed below these portions.

(b) The distorted portions of the bottom flange plates to be removed and new flange plates of same section to be fixed in their place.

In view of the nature and extent of the distortions, straightening of the flange plates *in-situ* within acceptable tolerances as envisaged in alternative (a) above was considered time consuming and possibly not practicable. This alternative was therefore discarded in favour of alternative (b) and rehabilitation was done accordingly. Figure 8.2 *(b)* and *(c)* show the rehabilitation scheme adopted.

Since the bridge has a very busy roadway below, it was not possible to stop traffic for providing supports to the plate girders from the roadway. Instead, the existing structure was supported from the top by connecting these to temporary girders placed over the existing ones.

(Source: Ghosh, UK & Mitra S: *Rehabilitation of Steel Bridges*, paper presented in 4th International Seminar on Bridges & Aqueducts 2000, organized by Indian Institution of Bridge Engineers, Bombay, 1998.)

8.4 RAKEWOOD VIADUCT, UK

Constructed during 1967–1969, Rakewood viaduct is a steel/concrete composite bridge carrying a dual three-lane motorway (M 62) across a 36 m deep valley adjacent to the Lancashire/Yorkshire boundary in England. It is 34.4 m wide six-span continuous structure of 256 m overall length, with four central spans of 45.7 m and two end spans of 36.6 m. The cast-*in-situ* deck slab was designed to act compositely with 3.05 m deep welded plate girders.

The viaduct lies in a section of the motorway where there is a steep gradient. Since the percentage of commercial vehicles using this motorway is high, there had been severe congestion in the carriageway where these heavy vehicles had to climb up the gradient, particularly at times of high winds and heavy snow and icing, causing the traffic to slow down. In order to reduce the problem it was decided to add a fourth climbing lane along with a new hard shoulder on the affected side of the carriageway. However, for the viaduct portion, instead of undertaking extensive widening work, it was decided to modify the layout of the lanes and accommodate the fourth climbing lane within the existing width of the viaduct by dispensing with the hard shoulder.

The structure (which was originally designed to BS-153) was checked as per the current BS-5400 design requirements and also BS-5400 loadings with modified lane layout (four-lane instead of three-lanes + hard shoulder). The check revealed that the girders were inadequate over the piers where the compression flanges (bottom flanges) were overloaded due to increased hogging moments. In order to reduce these hogging moments, it was decided to adopt external prestressing technique. The arrangement has been shown in Fig. 8.3.

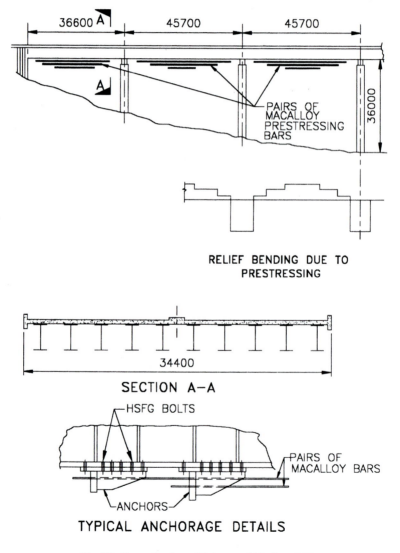

Fig. 8.3: Strengthening of Rakewood Viaduct, UK

First, fabricated steel anchors were fixed to the underside of the bottom flanges of the steel girders by HSFG bolts. Three pairs of Macalloy prestressing bars of overlapping lengths were then attached under each flange between the piers. When these bars were stressed, hogging bending moments were introduced in the mid-span region which, due to the continuity of the span, induced sagging moments over the piers. Thus the hogging moments at these locations (over the piers) due to load on the spans were reduced considerably and kept within the permissible limits. Additional stiffeners were welded to the webs at the locations of the anchors to ensure effective dispersion of the high anchorage loads from the anchors into the girder flanges and webs and thereby keep the local stresses within limits.

Needless to add, the system of strengthening adopted here involved minimal disruption of the traffic along the motorway, as the work was carried out mostly from the underside of the carriageway.

(Source: Pritchard, B: 'Bridge strenthening with minimum traffic disruption' *Proceedings of the Conference on Bridge Modification*, organized by the Institution of Civil Engineers, London. Thomas Tellford, London, 1995.)

8.5 STEEL/CONCRETE COMPOSITE BRIDGE IN NORTH CENTRAL IOWA, USA

A three-span continuous steel/concrete composite girder bridge in north-central Iowa was strengthened partially by attaching prestressing bars on top of the bottom flanges between the piers on either side of the webs. The arrangement was somewhat similar to that used in the strengthening done for the Rakewood Viaduct (described separately in this chapter), except that the bars in this case were fixed on top of the bottom flange within the depth of the girder.

The remainder of the strengthening was done by attaching a special truss mechanism over the piers on either side of the web and within its depth. The system adopted has been shown in Fig. 8.4. The truss mechanism was formed by a 30 mm diameter tendon acting as top chord of the truss with two diagonal members attached to the girders as shown. As tension was applied to the tendon, the upward reaction at the truss ends produced a sagging moment in the girder at the location of the piers, thereby relieving the remaining hogging moment at this location of the girder due to loading on the spans. The advantage of this mechanism of truss formation is that, it eliminates the high local stresses associated with the system of using

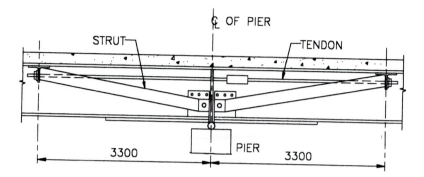

Fig. 8.4: Prestressed truss mechanism over pier for strengthening girders in north central Iowa, USA

anchors for prestressing, which normally requires addition of local web stiffeners.

(Source: Pritchard, B: 'Bridge strengthening with minimum traffic disruption' *Proceedings of the Conference on 'Bridge Modification'* organized by the Institution of Civil Engineers, London, Thomas Telford, London, 1995.)

8.6 BRIDGE OVER THE LITTLE CALAMET RIVER IN COOK COUNTY, ILLINOIS, USA

The 128 m span truss bridge over the Little Calamet river in Illinois suffered deterioration over the years. Addition of a heavy deck system resulted in considerable reduction of the live load capacity of the bridge. Since the dead load stresses were very high, it was found that addition of strengthening materials to members would not enhance the live load capacity significantly. Therefore a two-stage rehabilitation scheme was adopted. In the first stage the stringer beams of the deck system were replaced. In the second stage, selected members were post-tensioned by attaching prestressing strands at their ends, thereby inducing compression stresses in the tension members and tension stresses in the compression members. With this scheme dead load stresses in members were considerably off-set resulting in increase in the live load capacity of the bridge. Figure 8.5 shows a schematic diagram of the post-tensioning system adopted.

(Source: Swindlehurst, J and Parkinson, FH: 'Steel Structures', *Bridge Inspection, and Rehabilitation'*, Parsons Brinckerhoff, © John Wiley & Sons, Inc, USA, 1993. Reproduced by permission of John Wiley & Sons Inc.)

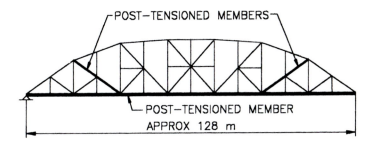

Fig. 8.5: Post tensioning arrangement for upgrading span over the Little Calamet river, USA

8.7 GOLDEN GATE BRIDGE IN SAN FRANCISCO, USA

This suspension bridge with a mainspan of approximately 1280 m was completed in 1937. The original 17.8 cm thick reinforced concrete deckslab carrying six 3.05 m wide traffic lanes was supported by rolled beam stringers placed at 1.45 m spacing. Extensive cracks in the reinforced concrete deck slab was noticed in the late 1960s. Inspection and tests revealed that the slab had not only separated from its supporting stringers at many locations, but also the concrete was highly contaminated by chloride and consequently needed to be replaced. Since the bridge provided the only arterial vehicular route between San Francisco and the counties to the north, it was necessary to adopt a modular pre-frabricated replacement system, in which construction operation could be carried out at site only during night time, thus allowing all the six traffic lanes to be in use during the day. Steel orthotropic deck system was found to be most suitable for the purpose and was finally adopted for the rehabilitation work.

The new orthotropic deck consists of a 16 mm thick steel plate stiffened by 28 cm deep, 10 mm thick longitudinal trapezoidal shaped ribs welded to the underside of the deck plate and supported on the main cross girders at stiffener locations by pedestals. For easy operation at site four pieces of modules—two each of 15.24 m length by 4.35 m width and 15.24 m length by 5.10 m width were used. These modules were pre-fabricated at workshops away from the site and brought to the yard at the site, where a first coat of pavement of crushed stone embeded in epoxy mastic (12.7 mm thick) was applied. These were then erected and supported on the cross girders. While the actual replacement work was carried out at night only, all preliminary work necessary for removal of the existing concrete slab and its supporting steel stringers was carried out during the day time from a platform suspended under the deck. Thus all the six lanes of the

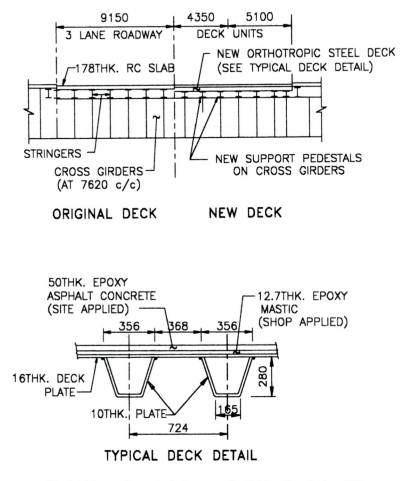

Fig. 8.6: New orthotropic deck system for Golden Gate Bridge, USA

bridge were available for use during day time for traffic throughout the period of construction. With this system the complete redecking from the first deck unit to the last took 401 working days. The final wearing surface of 5 cm thick epoxy-fortified asphalt concrete was placed after completion of the deck erection. Figure 8.6 shows a schematic diagram of the system adopted.

(Source: Stahl, Frank L: 'Orthotropic steel plates for bridge deck replacement' *Extending the Life of Bridges*, ASTM STP 1100 GW Maupin Jr., BC Brown and AG Lichtenstein, Eds., American Society for Testing and Materials, Philadelphia, 1990.)

8.8 GEORGE WASHINGTON BRIDGE NEW YORK, USA

Opened to traffic in 1931, George Washington Bridge across the Hudson river in New York is a suspension bridge with a mainspan of approximately 1067 m and has two roadway decks. The original deck slab of the upper deck carrying two 13.4 m wide four-lane roadways consisted of 21.6 cm thick reinforced concrete slab spanning between secondary floor beams placed at 1.57 m spacing. Deterioration of this deck became noticeable in the late 1950s and continued at an accelerated pace needing nagging patch-repair work at decreasing intervals. This was causing considerable inconvenience to the commuting traffic on this important and busy crossing. It was therefore decided to replace the concrete deck by an orthotropic steel deck system and the work was carried out in 1978–79.

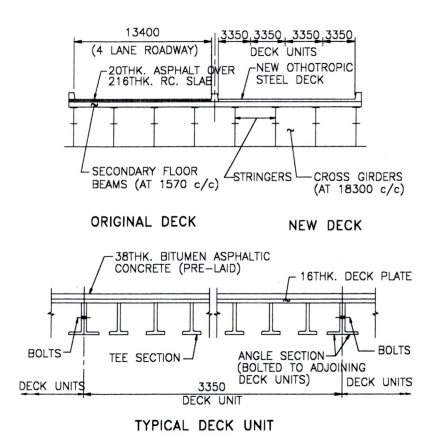

Fig. 8.7: New orthotropic deck system for George Washington Bridge, USA

The new orthotropic deck consists of a 16 mm thick weathering steel deckplate stiffened by tee ribs welded to the underside of the deck plate and fixed on the existing secondary floor beams by high strength bolts. Bitumen asphaltic concrete pavement of 3.8 cm thickness was pre-laid on the deck plate to reduce operation time at site. The replacement work was carried out during night only—replacing each night a section of 13.4 m width by 18.3 m length, i.e. one of the two four-lane existing roadways between two floor cross girders placed at 18.3 m centre. To suit easy site operation four pieces of new deck modules of 18.3m length by 3.35 m width were prefabricated and brought to the site along with the already applied pavement on the top. These were then erected side-by-side and bolted together along their longitudinal edge to make up the total width of 13.4 m. Since the pavements were already applied at shops, the section would be ready immediately after erection, making all the eight lanes available for the morning rush hour traffic. During off-peak hours and at night all the traffic was diverted to one of the two four-lane roadways, leaving the other roadway free for carrying out preparatory activities for actual replacement work to be done during night time. With this system the entire work was carried out well within the specified time frame. Figure 8.7 shows a schematic diagram of the system adopted.

(Source: Stahl, Frank L: 'Orthotropic steel plates for bridge deck replacement', *Extending the Life of Bridges*, ASTM STP 1100, GW Maupin Jr., BC Brown and AG Lichtenstein, Eds., American Society for Testing and Materials, Philadelphia, 1990.)

8.9 THROGS NECK BRIDGE APPROACHES, NEW YORK, USA

The approach Viaducts to the Throgs Neck Bridge in New York City comprise of a number of simply supported twin girders, of spans ranging between 42.7 m and 57.9 m, carrying two 11.6 m wide three-lane roadways, separated by a median divider. The original 19 cm thick reinforced concrete deck slab was supported by rolled beam stringers spanning between cross girders with cantilevered projections on two sides beyond the main plate girders. These cantilevered portions of the cross girders had to carry the outer truck lane as well as a portion of the middle lane. Extensive cracks in the concrete deck was noticed in the outer lanes in early 1970s. Gradually the cracks progressed in the other lanes also causing concern to the authorities.

Inspection was carried out followed by design check. This revealed that the original deck slab was designed with rigid supports at stringers. The comparatively shallow and flexible cross girder cantilevers on which the stringers were supported could not provide the desired rigidity considered in the design. This resulted in differential deflections in the stringers, which induced additional stresses in the slab leading to the cracks. The distress as well as chloride contamination from de-icing salts in the slab prompted the authorities to consider replacement of the deck slabs. Since it was imperative to keep the traffic moving in all the six lanes during day time, it was decided to replace the concrete decks by othotropic steel deck system, for which construction work would be carried out in the nights only.

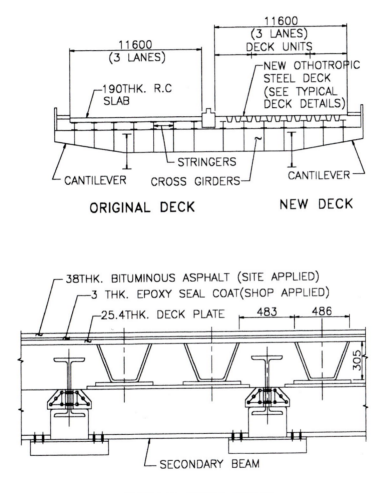

TYPICAL DECK DETAIL

Fig. 8.8: New orthotropic deck system for Throgs Neck Bridge approaches, USA

The new orthotropic deck consists of 25.4 mm thick steel plate stiffened by 30.5 cm deep, 11 mm thick longitudinal trapezoidal ribs welded to the underside of the deck plate and supported on secondary cross beams spanning between the stringers. The orthotropic plate system is continuous within each span of the approach viaduct. The transverse and longitudinal deck plate splices are field-welded, and the rib splices are bolted. A 3 mm thick epoxy and grit seal coat was pre-laid in the shop which was topped later in the field with a 38 mm thick nylon fibre fortified bituminous asphalt final course. Deck panels were of different lengths (up to 15.8 m) to suit the span lengths of the main girders. The width of the panels varied between 3.66 m and 3.96 m. During each night one of the roadways was closed and the traffic was diverted to the other roadway. Sections of the existing slab were removed for a full width of the roadway each night and new deck panels erected in its place. In the morning all the six lanes were again opened to traffic causing minimal hindrance to the peak period rush. Figure 8.8 illustrates the scheme adopted.

(Source: Stahl, Frank L.: 'Orthotropic steel plates for bridge deck replacement', *Extending the Life of Bridges*, ASTM STP 1100, GW Maupin Jr., BC Brown and AG Lichtestein, Eds., American Society for Testing and Materials, Philadelphia, 1990.)

8.10 BENJAMIN FRANKLIN BRIDGE, PHILADELPHIA, USA

Opened to traffic in 1927, Benjamin Franklin Bridge over the Delaware river in Philadelphia is a suspension bridge with a main span of 533 m. The original deck slab carrying a 23.7 m wide seven lane roadway consisted of a 16.5 cm thick reinforced concrete slab spanning between longitudinal stringers. The deck slab was found deteriorating due to heavy traffic and chloride contamination from de-icing salts. The steel stringers were also found corroding due to water travelling through the deck joints. It was therefore decided in 1982 to replace the original damaged concrete slab as also the steel stringers by an orthotropic steel deck system.

The new orthotropic deck system consist of a 16 mm thick deck plate stiffened by specially rolled 31.8 cm deep bulb sections welded to the underside of the deck plate and supported on the existing cross girders. The deck units are made continuous. This system enables the new deck not only to act along with the stiffening truss system, but also to contribute to the flexural and torsional rigidity of the structure and to improve its aerodynamic characteristics. Field splices of deck plates and ribs are connected by bolts. Epoxy asphalt of 3.2 cm thickness was pre-laid in the shop to reduce operation time at the site. The final wearing surface of 32 cm thick bituminous asphalt was placed after completion of erection of the deck.

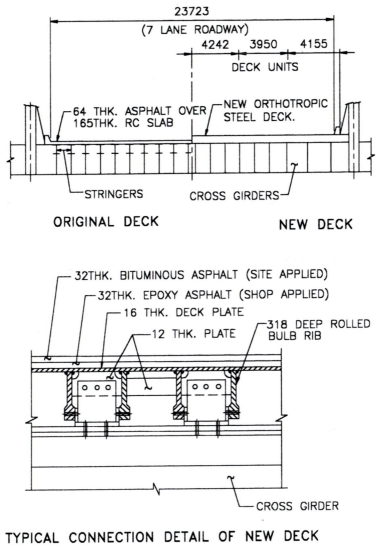

Fig. 8.9: New orthotropic deck system for Benjamin Franklin Bridge, USA

The entire construction work was carried out in four phases in the day time only. During each phase traffic blocks were imposed in two lanes in peak hours and three lanes in off-peak hours. Figure 8.9 shows a schematic diagram of the system adopted.

(Source: Stahl Frank L.: 'Orthotropic steel plates for bridge deck replacement', *Extending the Life of Bridges*. ASTM STP 1100, GW Maupin Jr., BC Brown and AG Lichtenstein, Eds., American Society for Testing and Materials, Philadelphia, 1990.)

8.11 CASTLEFIELD BRIDGE, MANCHESTER, UK

Built in 1906, this Warren truss bridge formed a part of the restoration of the Cornbrook viaduct for the Light Rapid Transport system in Manchester, which started in 1989. The bridge crosses an existing British Rail track below at a 45 degree skew.

Overhead electrification of this railway line was being contemplated about that time needing additional clearance at the bridge location. The existing Castlefield Bridge was thus required to be raised for the electrification project. At the same time, it was found that the roller bearings of the bridge needed to be replaced. The jacking operation for this replacement

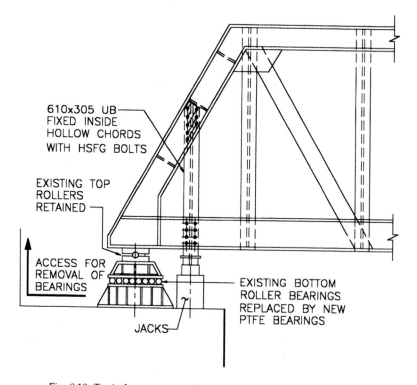

Fig. 8.10: Typical arrangement for jacking of Castlefield Bridge, UK

work provided an opportunity to raise the bridge permanently to give the required clearance for the electrification project and also to improve the vertical alignment of the new LRT track.

The scheme envisaged raising of the bridge by about 150 mm and replacing the existing roller bearings by new PTFE/stainless steel sliding guided bearings, retaining the existing top transverse rollers to allow longitudinal rotation. Since no space close to the intersection of the bottom chord and end raker was available for jacking, it was proposed to provide an alternative system at a location between the intersection node point and the first vertical. At this location Universal Beam jacking posts were inserted inside the hollow bottom chords and end rakers of the two trusses and secured by HSFG Bolts (Fig. 8.10). New concrete pedestals to suit the level of the new PTFE bearing was done by raising the entire 420 tonne bridge during night time.

(Source: Mathews RA and Patterson IA, 'LMR to LRT: Restoration of the Cornbrook Viaduct', *Bridge Management-2: Inspection, Management, Assessment and Repair*, Edited by JE Hardinge, GAR Parke and MJ Ryall, Thomas Telford Limited, London, 1993.)

8.12 STEEL/CONCRETE COMPOSITE GIRDERS OF THE LONDON DOCKLAND LIGHT RAILWAY, UK

Completed in 1987, the new viaduct of the London Dockland Light Railway consists primarily of twin steel girders supporting cast-*in-situ* reinforced concrete deck slab. The original design to BS-5400 envisaged the concrete deck to act compositely with the welded plate girders for which shear connectors were welded on the top flange of the girders. However, after completion of the construction, the decision to build the Canary Wharf Development resulted in substantial increases in the weight and frequency of trains with consequent reduction of fatigue life of the deck welded shear connectors (in some cases as much as 75%). In order to increase the fatigue life of these connectors to 120 years as envisaged in the original design, it was found necessary to install additional shear connectors between the original welded steel connectors.

To minimise traffic disruption, the possibility of installing the new shear connectors from under the top flange of the steel girders was examined. Several types of connectors were considered. Out of these, 'Spirol' spring pin shear connectors were found to be the most suitable, as these steel fasteners offered the advantage of a readily achieved force-fit into the hole drilled through the steel flange and the concrete deck slab with no requirement of grouting, glueing or welding. In this system the pin with an in-built spring mechanism of spirally coiled steel strip is force-fitted into the drilled holes by jacking. The spirally coiled pin then induces spring loaded friction between the pin and the hole. In the viaduct in question the steel spring pins were successfully installed from the underside of the girders without interrupting traffic flow.

(Source: Pritchard, B: 'Bridge strengthening with minimum traffic disruption', *Proceedings of the Conference on 'Bridge Modification'* organized by the Institution of Civil Engineers, London, Thomas Telford, London, 1995.)

8.13 HARDINGE BRIDGE ACROSS THE GANGES, BANGLADESH

This elegant bridge across the Lower Ganga is situated near Paksey in Bangladesh and consists of fifteen 105 m Petit type riveted through spans and six 23 m deck type riveted approach spans. Opened to traffic in 1915 the bridge carries double line broad gauge (1,680 mm) railway traffic. The bridge suffered serious damage during the brief war in the Indian sub-continent in December 1971. The bottom chord and diagonal members of the ninth span from the western bank was damaged by explosion, which blew off approximately 18 m of the length of the bottom chord in the central region of the downstream truss along with several web members of the truss in that region. Furthermore, the deck system in the central panel was substantially damaged. The downstream truss sagged by 105 mm, whereas the upstream truss sagged by 25 mm. The span also tilted out of plumb by a slope of 1 in 250 at the centre towards the downstream side. Although the individual truss in this condition became theoretically unstable by planer analysis, total collapse apparently was prevented by the presence of the extensive system of the top and bottom lateral bracing along with sway bracings at every cross frame locations, which induced the structure to act as a space frame. The damaged span was analysed as a space frame by a computer. The analysis led to the conclusion that none of the undamged members had reached yield stress. Therefore the rehabilitation work was to be limited to the damaged components only.

The rehabilitation work was of a challenging nature, considering the unfavourable river condition as also the economically strategic location of the bridge, which implied that the bridge was to be recommissioned with the least possible delay.

Rehabilitation work of the damaged members of the span was carried out after temporarily supporting the damaged downstream truss from adjacent downstream trusses with the help of pin connected link members. For this purpose the truss was jacked up by adopting barge floatation principle, by employing two steel barges with water tight bulk heads, which

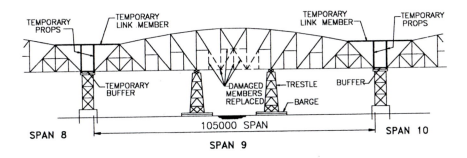

Fig. 8.11: Rehabilitation of Hardinge Bridge, Bangladesh

were specially modified and strengthened for this purpose. Prior to jacking up operation the barges were filled with water and the truss was supported at two nodal points of the bottom chord by means of steel trestles fixed on the barges. As water was pumped out of the two barges simultaneously, equivalent jacking force was transferred on to the two nodal points of the bottom chords of the truss. With this upward thrust, the span moved back to its original position prior to the damage. In this condition the specially designed fabricated steel links were connected between the ends of the top chord of the downstream girder of the damaged span with the corresponding top chord ends of the adjacent trusses. Buffers were also fixed between the ends of the corresponding bottom chords to counterbalance the forces due to possible cantilever behaviour of the damaged truss, in case it detached into two separate parts. After the introduction of link and buffer arrangement, removal and repair work of members were undertaken without any danger of instability in the span. New members for replacing damaged members were fabricated in advance in workshops in Calcutta, making suitable adjustments in the lengths for accommodating the dead load stress condition of the existing truss.

Figure 8.11 shows the arrangement of the jacking up for rehabilitation work.

(Source: Ghoshal A, Ganguly JC, Banerjee HK and Kapoor MP: 'Hardinge Bridge Span Repair', *Journal of Construction Division*, American Society of Civil Engineers, December, 1974.)

8.14 ROAD BRIDGE OVER RIVER BICHOM, ARUNACHAL PRADESH, INDIA

Situated at the foothill areas of the Himalayas, the original Bichom bridge was a 68 m long single span through type steel truss bridge with R.C. road

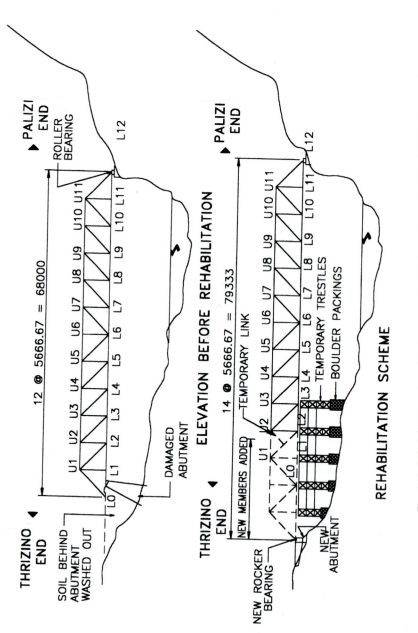

Fig. 8.12: Rehabilitation of road bridge over river Bichom, Arunachal Pradesh, India

deck. It was designed to carry class 40 R live load as per Indian Road Congress (IRC) standard. In June 1993, a severe and sudden flash flood washed out the soil behind one of the abutments and caused the abutment to tilt towards the river. As a result the span was resting precariously on the tilted abutment at the end panels of the bottom chords at some distance from the end nodal point—the load being transferred through the end panels of the bottom chords. Consequently the bottom chords at the end panels were deformed badly.

Since the bridge was of vital importance for communication in the strategic area, it was necessary to restore the link at the earliest possible time. To meet this requirement a simple and comparatively easy-to-execute solution was called for. To avoid recurrence of similar calamity in the future, it was considered necessary to increase the waterway and construct a new abutment at some distance behind the damaged abutment. To cater for this requirement, the rehabilitation scheme was prepared envisaging a longer span superstructure. Since construction of a new superstructure would take considerable time, it was decided to retain the undamaged portion of the existing bridge with minimal modification and increase the span to 79.3 m by adding two panels.

Studies carried out for different load conditions revealed that the main members of the existing bridge structure would require major strengthening and repair work by way of replacement of main members, in order to rehabilitate the bridge to its original capacity. The span was also checked for a condition with new orthotropic steel plates replacing the existing R.C. deck. It was found that while the truss members were adequate with Class 24R IRC loading, the deck supporting members (cross girders and stringers) were inadequate, as these were earlier designed as steel-concrete composite sections. Rehabilitation of the entire deck system would therefore be a time consuming process. Since it was considered necessary to restore the link at the earliest possible time, the possibility of rehabilitation with even a lower live loading was examined. Thus, the superstructure was checked with Class 18R IRC loading and was found to be adequate with only minor strengthening of some of the existing members and without the necessity of dismantling of any member of the existing portion proposed to be retained. This alternative was finally adopted. The stages of rehabilitation has been shown in Fig. 8.12.

(Source: Ghosh UK, Das PP and Ghoshal A: 'Rehabilitation of Steel Bridges—Experiences in India and Malaysia', *The Bridge & Structural Engineer*, Indian National Group of the International Association for Bridge & Structural Engineering, New Delhi, March 1997.)

Safety During Field Work

9.1 INTRODUCTION

In a bridge rehabilitation programme, field activities are involved primarily on two occasions viz., during inspection and during actual rehabilitation work. Inspection activities are normally limited to short spells and do not require heavy or bulky equipment. On the other hand, rehabilitation work is, in most cases, a comparatively long process and may require heavy equipment such as cranes, compressors, etc. During both these operations, safety of the users of the bridge and of the local public assume prime importance. No less important is the safety of the inspection and repair crew and this must also be adequately addressed in this context. Since inspection and rehabilitation work are carried out when a bridge is in service, safety aspect becomes closely related with the arrangements for controlling the traffic during these operations. Traffic control, therefore, needs careful planning to ensure adequate safety during inspection and rehabilitation work.

9.2 SAFETY ARRANGEMENTS

Many authorities such as Railways and Public Works Departments have their own safety rules and regulations including those for special items like railway tracks, electrified lines, etc. They normally stipulate these rules for works within their reserve. Members of the working team must be aware of and comply with these safety rules. Most of these regulations however, are meant for their departmental use for normal day-to-day working and may not address the specific requirements of rehabilitation work. Some typical safety aspects associated particularly with rehabilitation work are therefore discussed in the following paragraphs.

9.2.1 Temporary Construction Screens

In cases where roadway or railway movements are allowed during the rehabilitation work, the necessity of protecting the traffic becomes important.

Temporary construction screens are to be erected to isolate the construction works from the areas being used for vehicular or pedestrian movements. Whenever overhead works are carried out, moveable overhead screens to protect against falling items such as tools, loose bolts/nuts, debris, etc., should be erected. Pedestrian enclosure with overhead cover gives better protection to pedestrians than that given by simple barrier without overhead cover. The type of construction screens to be adopted in a rehabilitation work should be carefully considered during the design of the scheme and should best be incorporated in the contract documents.

9.2.2 Proximity of High-voltage Power System

Extreme caution should be exercised during rehabilitation work of a bridge where high voltage traction systems such as third rail or overhead catenary are used. In such cases power shut off or adequate protective shielding must be enforced prior to commencement of any inspection or repair work. Also, the safety regulations required by the railways in respect of electrified tracks must be followed.

9.2.3 Movement Across Railway Track

In case of rail bridges, plant and construction materials may be required to be shifted from one side of the track to the other. This should be done in close coordination with the railway authorities and with their prior permission and preferably under the supervision of authorised railway officials. Since trains normally operate at high speeds and need a longer distance to decelerate and come to a half, it is imperative to have experienced railway staff for flagging during site work. This would reduce the chances of accidents considerably.

9.2.4 Clearances

Allowable clearance envelope for traffic may need some temporary infringement during inspection or rehabilitation work. This may be due to erection of temporary platforms or rigging requirements, or employment of snoopers to approach the underside of a bridge deck. This may be particularly significant in a bridge with roadways or railway tracks underneath, or where a bridge crosses navigational waterway. Safety aspect arising out of such reduced clearance envelope should be discussed with the concerned authorities well in advance of the proposed date of inspection or repair-work.

9.2.5 Public Utilities

Many bridges carry facilities of public utilities such as power, gas, water or oil pipelines, etc. as also wires for railway signalling. All these facilities are of vital importance and the repair crew must be made aware of this. Under no circumstances should these be disturbed unilaterally.

9.2.6 Weights of Equipment

Construction equipment can impose significant loads on bridge structures, particularly the loads concentrated at the outriggers of mobile cranes or snoopers. Local effect of such loads should be checked before finalizing the equipment to be used in the work.

9.3 TRAFFIC CONTROL FOR SAFETY

Traffic control for safety is generally influenced by two major factors viz., the type and volume of traffic to be controlled and the proposed planning for site activities.

9.3.1 Traffic

The word 'traffic' should include not only vehicular traffic like cars, buses, trucks, etc., but also the pedestrian flow over a bridge. The types of traffic and their volumes passing over a bridge is usually influenced by its distinctive characteristic which depends on a number of factors. Thus, an aesthetically pleasing bridge set in a natural environment or a bridge built with modern technology is likely to attract tourists or persons interested in technological excellence. Light vehicles or coaches are likely to form the bulk of the traffic on such a bridge. On the other hand, a bridge connecting two commercial or industrial centres would attract heavier types of vehicles. A bridge of strategic military importance, may have completely different types of traffic. Also the width and number of lanes have a bearing on the volume of traffic in a bridge. Thus, a four-lane bridge is likely to carry smaller traffic than a wider bridge with eight lanes. For a bridge located near a busy urban conglomerate, the likely peak volume would be during the morning office rush hour, the mid-day lunch period and the evening hours after the offices and business establishments close. Also occasional pedestrian assemblage may occur in a bridge located near a stadium or similar facility. Traffic control plan must address all such possible situations. Thus, it becomes almost imperative that the traffic control plan has to be done specifically for each bridge, considering the anticipated traffic flow over it. Needless to add that prior knowledge of the type and volume of such traffic is always desirable for developing an effective and safety oriented traffic control plan.

For road bridges the traffic data are likely to be available with the concerned authorities. Otherwise traffic survey may have to be carried out, particularly for busy and strategic locations.

Traffic control requirements for railway bridges need some special considerations. Railway services generally operate on pre-arranged time schedule. It is, therefore, possible to know beforehand the periods when the bridge is not likely to be used and programme repairwork accordingly. However, if the nature of repairwork is such as not to match this time period, it may be necessary to close the line temporarily for a specified period and carryout the repairwork. Since rescheduling of rail movement is likely to have wide ranging side effects, it should be avoided, or at least limited to the minimum. In a multi-track system, one line in a bridge may be closed for carrying out repairwork in its vicinity only, while allowing rail movements on the other lines. Safety aspects in such cases need special examination. During any rehabilitation work, it is always advisable to keep experienced railway staff at the site for alerting incoming trains as well as the repair crew.

9.3.2 Planning for Site Activities

Ideally a plan for site activities should precede the traffic management plan. Proposed methods of access for inspection or rehabilitation work, details of erection scheme, types of equipment to be used, number of skilled and unskilled labour, duration of site activities, etc., are vital factors for developing a safety oriented traffic management plan. If required, these may even need modifications to suit safety requirements. Thus, plannings of site activities and traffic management should be finalised by the concerned authorities in close coordination with each other. A schedule for traffic blocks for carrying out site activities, which require stoppage of normal traffic, needs to be drawn beforehand. If the duration of such a block is arranged to be long enough to carryout all activities requiring traffic stoppage, it is always preferable to complete all such work first and do the non-traffic affecting activities later on. These agreed plannings may, however, need further modification and updating as the work at site progresses, since duration and sequence of activities may change due to various unforeseen reasons such as equipment failure, weather conditions, etc.

9.3.3 Traffic Control Methods

Sufficient precautions should be taken to ensure safety of all traffic including pedestrians and also the inspection/repair crew in and around the worksite. There are a number of methods, normally used for this purpose.

(a) Posting of Traffic Signs

Generally the speed limit for vehicles near the worksite is less than the posted speed limit. It is, therefore, necessary to display signs specifying the lower speed limits at sufficient distance from the work site. Signs cautioning drivers about repair work being done on the bridge should also be displayed prominently well ahead of the worksite.

(b) Barricades and Channelising Devices

Standard channelising devices such as drums, cones, fabricated removable barricades are generally used for keeping the traffic at a safe distance from the work zone.

(c) Warning Lights

Portable flashing lights placed at strategic locations are very useful for drivers even during day time and provide additional safety to the workmen. Barricades which are required to be left in place overnight should have adequate lighting arrangement as warning to the drivers passing over the bridge.

(d) Flaggers

In a two-lane two-direction bridge, where one lane is closed for inspection or repairwork, generally two men are placed at the ends of the bridge, armed with flags for controlling the traffic to move alternately in the two directions. Traffic lights may also be used to serve the same purpose.

9.4 CONCLUDING REMARKS

The importance of safety during inspection and rehabilitation work at site cannot be overemphasised. However, this aspect normally does not get the due attention that it deserves. The main reason for this attitude is perhaps the fact that repair and rehabilitation work does not enjoy the charisma of a new construction work. Nevertheless, brief outline of the subject has been discussed in the preceding sections. However, as mentioned earlier in this chapter, each bridge has its own individual characteristic and therefore it is not possible to perceive in this general discussion every safety aspect for a particular bridge. Since nearly every bridge in existence today has undergone some repairs in the past as part of normal maintenance work, it may be worthwhile to discuss safety aspects with persons who are or were connected with the maintenance of the particular bridge and are familiar with the site conditions and utilise their experience in the matter. Ensuring the safety of all concerned must be the topmost priority in any rehabilitation project.

Bibliography

1   *Bridge Inspection Guide*, Her Majesty's Stationery Office, London, 1983.
2   IS: 7205–1974 *Safety Code for Erection of Structural Steelwork*, Indian Standards Institution, New Delhi.

Post Rehabilitation Maintenance Guidelines

10.1 INTRODUCTION

Bridge Structures are constantly exposed to environmental and various other adverse effects all round the year. These may, additionally be subjected to intense industrial pollution and unfavourable marine situations. It is therefore necessary to protect a rehabilitated bridge from deterioration due to such effects and ensure that the investment made for rehabilitation is not frittered away prematurely. It is heartening to note that, of late, attitude towards preservation of our existing assets is changing and the importance of introduction of a well-planned and monitored inspection and maintenance system for existing bridges is being increasingly appreciated. Such a system would ensure periodic inspection and recording of the current state of the structure, thereby providing feedback information about the actual and potential sources of danger for taking timely remedial actions.

Inspection procedure for determining the physical conditions of various members and other related aspects such as selection of personnel, areas to be inspected, tools, equipment, testing methods, instrumentation, etc. have been discussed in Chapter 3. Similarly repair work to damages due to corrosion and other reasons have been extensively discussed in the previous chapters. These aspects are therefore not repeated here. This chapter is intended primarily to provide guidelines for preventive maintenance activities needed for a rehabilitated steel bridge. While each situation requires careful thought and judgement, these guidelines would be helpful in achieving the overall objective of adequate maintenance of the structure.

10.2 DOCUMENTATION

The first requirement for an effective maintenance system is to prepare a methodically entered current record for each bridge under the system.

These records may be collectively stored in a computerised data bank. The following basic information should be available from the record:

- Number or name of the bridge
- Year of construction
- Location giving distance from the nearest town, village, etc. A small scale map indicating the position would be very useful.
- Type of the structure, e.g. plate girder, type of truss, rivetted or welded, etc.
- General dimensions including overall length, width, number and length of spans, skew angle, if any, depth and width of girders, etc.
- Services carried across the bridge
- Details or types of components such as bearings, expansion joints, parapets, guard rails, etc.

The recordings should also indicate clearly the present live load capacity of the bridge supported by design calculations.

The records should provide full history of each structure, including drawings and details of repair or rehabilitation work undertaken previously and the behaviour of the structure thereafter. The recordings should be preserved systematically in a concise format, so that information for a particular bridge can be easily retrieved. Such retrieved information would be very useful as a background for an ensuing inspection, particularly if there are special problems needing prior discussions with the concerned engineer. Important features arising out of such inspections should in turn, be stored in the records for future reference.

10.3 TRAINING

Co-ordinated participation of a team of experts is the essence of an effective inspection system. Training programmes are needed in order to develop such expertise. Facilities for training should be set up at central level as well as at local levels for training field staff. A continuing training programme at regular intervals will help the grassroot workers to update their technical expertise in their respective fields.

10.4 FREQUENCY OF INSPECTION

The frequency of inspection and the extent to which a bridge is to be inspected depends on many factors such as age, condition of the structure, traffic characteristic, strategic importance, environment, history of deteriorating conditions, etc. The person in charge of the inspection programme of a particular bridge should consider the above factors prior to deciding on the frequency of inspection. As a guideline, each bridge should be subjected to two types of routine inspection, viz., General Inspection and Principal Inspection. General Inspections are made on certain items at intervals of say 1 year by visual observations from ground or deck level or from existing walkways or platforms. Principal Inspection, on the other hand, is detailed inspection of all parts of the bridge needing suitable temporary arrangements for access. The latter type of inspection may be carried out say once every five years. Special Inspections are required for a bridge when a particular problem needs investigation, such as damage due to accident, flood, loss of camber, etc. These are undertaken on the basis of reports of unusual occurrences or routine inspection reports.

10.5 MAINTENANCE REPAINTING

10.5.1 General

Primarily maintenance re-painting of a steel bridge is done to protect it against corrosion, so that the structure can perform its intended functions over a specified period of time without the need of structural repair. The other function, though less important, is to preserve the overall appearance of the bridge. Often, the latter function is given more importance, with the assumptions that once the appearance aspect is satisfied, the steel would be automatically protected. However, this assumption may not necessarily be correct. Some locations not easily visible are likely to escape the attention of the viewer, although these parts may not have been painted properly and may have been exposed to corrosion. It is therefore necessary that the primary objective for maintenance re-painting, namely protection of the steel is not undervalued.

10.5.2 Common Problems of Re-painting

Many of the existing steel bridges around the world receive coats of protective paints at regular intervals costing the owner large sums of money. It has, however, been noticed that quite often, the location which are easily accessible are painted regularly, whereas other areas which are not so easily accessible do not receive proper attention. As a result the areas which have easy access are generally not rusted over the years. On the other hand, in the same bridge, parts which are comparatively less accessible have not been painted properly and consequently have suffered damage due to corrosion.

There have also been instances where fresh coats of paint have been applied without either removing all the rust or cleaning the area properly. As a result rusts have reappeared within a short period of time after re-painting. It is therefore necessary to pay special attention to cleaning such areas properly, to remove rust and other contaminations before applying fresh coat of paint. One point needs special mention here. In case the loss of sectional area is beyond the permissible limit, the member may have to be strengthened by adding corrosion plate prior to painting.

There is another aspect which needs looking into, viz., excessive number of coatings, on a member. While thick coatings may appear to provide more protection to the steel surface, it may, in fact be detrimental and may even induce cracking and flaking, needing removal of the entire coating.

In a steel bridge there are some parts where corrosion does not affect the overall stability of the structure, whereas in some members damage due to corrosion may be critical. It is therefore necessary to formulate strategy for identifying the more structurally sensitive areas and to introduce a monitoring system to ensure that damage due to corrosion in such areas are attended to promptly.

Efficiency of maintenance paintings of an existing bridge depends largely on the initial painting system adopted and the quality of maintenance painting done over the subsequent period. If the original painting system was inadequate for the service conditions, or the workmanship was not up to the desired level, it becomes difficult to achieve an efficient re-painting work. Similarly, long spells of neglect or inadequate maintenance may pose unusual problems to subsequent maintenance work, needing extensive cleaning and often repair work, and entailing large scale patch painting prior to final coats of painting.

10.5.3 Types of Coating Defects

Some of the common types of coating defects observed on existing bridges are briefly discussed in the following paragraphs:

(a) Blistering

Generally blisterings can be classified into two categories. The first type occurs within the coating itself caused by the pressure of water or solvents trapped within or under the paint film. The other type is caused by corrosion of the underlying steel base arising from soluble iron corrosion products at the interface of metal and paint. If the adhesion between the metal and the paint film is not adequate, the blistering is likely to spread.

(b) Flaking

Flaking is identified when a film of top coat peels off from the underlying coat. The main reason of this condition is loss of adhesive strength. Some of the causes of poor adhesive strength are:

- Presence of rust, dirt, millscale, powdery materials, etc. on the surface.
- Presence of contamination such as grease, oil, silicones, etc. preventing proper bond between the paint and the surface.
- Too smooth a surface for developing bond with the paint.

(c) Cracks

Cracks are not always visible without a magnifying glass. Surface hair cracks may be limited to the top coat only. However, some cracks penetrate beyond the top coat and ultimately cause failure of the protective system. Cracks are generally caused due to stresses on the paint films. The capacity of a paint against cracking depends on a number of factors, such as its condition during application and curing, plasticisation, age etc.

(d) Blooming or Blushing

This defect is common in regions with occasional heavy rains and is identified as local loss of gloss or dullness of colour (on the otherwise good paint surface) caused by condensation or high humidity during curing. This defect may reduce adhesion of subsequent coats. These can be removed by application of an appropriate solvent.

(e) Pinholes

These are minute holes in the film caused by bursting of air bubbles trapped in the paint during application, which the adjoining wet paint is unable to cover before the paint has set.

10.5.4 Selection of Protective System

It is a normal practice in bridges to apply the same type of protective paint as the existing one, provided its performance has been satisfactory. There may, however, be instances where the painting system needs upgrading because of inadequacy of the existing system, due to environmental change or other reasons.

Prior to selecting the type of new protective system, it is necessary to consider the following aspects:

(a) Compatibility

The main criterion for the selection of a new protective system is the compatibility of the material of the proposed system to the existing coating on the structure. Otherwise the new coat is not likely to adhere to the existing coat for a long period. In any case even for compatible systems, if the existing coat has hardened during service, it may need abrasive treatment for providing uneven keys on the surface to hold the new maintenance coat.

(b) Environment
The following aspects need consideration:

- Environmental conditions such as polluted or non-polluted, saline or non-saline
- Local conditions such as intermittent splash of water or fully immersed in water, presence of harmful salts such as sulphates or chlorides, likelihood of impact or abrasion, presence of fungi or bacteria.

(c) Overall Economy
The life-cycle costs for some prima-facie suitable systems should be compared.

(d) Other Factors
- Appearance (after final paint) *vis-a-vis* the surroundings
- Time required to provide the coating system
- Tolerance in surface preparation and different application treatments
- Suitability from considerations such as: availability, application methods (brushing, spraying, etc.), cost effectiveness, time schedule between the first and final treatment, etc.
- Past performance
- Level of expertise required in application of the coating system.

Selection of appropriate coating system with correct specifications are not the only criterion for satisfactory performance of the system. Failures can arise from incorrect use of materials that are basically sound. Thus application methods and tests are just as important as the quality of the material. Some of the sources of deficiencies are:

- Use of material after the expiry of shelf life
- Use of two-pack material beyond their pot life
- Use of wrong or excessive thinners
- Inadequate mixing of two-pack materials

10.5.5 Surface Preparation Prior to Re-painting

In any system, surface preparation is very important to make the new painting system effective. If the surface is not properly cleaned and not free from rusts or other chemicals, corrosion under new coating is likely to start again.

This has been the main problem for maintenance of a large number of existing bridges.

Maintenance painting is best done before any deterioration in the existing coating or corrosion in the base metal has taken place. This situation however, seldom occurs since maintenance paintings are generally done at predetermined intervals depending on the service condition. Minor damages noticed during the intervals have to be taken care of as a part of day-to-day maintenance work.

Generally the situations in localised areas which are required to be dealt with during maintenance painting operation of a bridge are as follows:

(a)   Where the existing system does not show any breakdown such as blistering or flaking, the surface will not require any special treatment. Maintenance painting may be applied after the surface has been thoroughly washed.

(b)   If the existing system is reasonably intact and shows only minor blistering or flaking locally with no rusting, the loose paint coatings are to be first removed followed by washing with clean water. Maintenance painting may then be applied after the surface has dried.

(c)   Where some areas show major damage of the painting system with or without rusting of the steel, it will be necessary to clean the steel thoroughly by power tools or flame cleaning or other appropriate methods to remove all the rust from the steel, in order to provide a firm base for the future coat. The full protective system i.e. priming coat and finishing coats are to be applied on the cleaned surface to complete the repair work. The surface is now ready to receive maintenance paint.

10.5.6 Financial Evaluation

The traditional practice for the financial evaluation of a protection system for steel bridge is to base the decision solely on the initial costs of different systems with scant regard to future maintenance costs. It is, however, increasingly being felt that this evaluation should be done on an economic basis, which considers long term effects of the different options. Life Cycle Costing (LCC) can fulfil this objective. The philosophy behind LCC is that, not only the initial costs but also all future maintenance costs to be incurred during the subsequent life span of the structure are to be considered for evaluating the total cost. Thus in an aggressive environment a special corrosion resistant painting system, which would initially cost more but would last longer than normal painting system may prove to be more economic over a period of time if analysed with LCC. Additionally, the life of the structure may be extended, thereby delaying heavy capital expenditure for replacement of the structure.

In this system the future costs are brought together into a single value, often by applying the traditional present value concept on a base date corresponding to the initial costs.

In the simplest form, present value can be calculated from the following formula:

$$\text{Present value } (PV) = \frac{C}{(1 + r)^t}$$

where C is the future costs at current prices, r is the discount rate expressed as a decimal, and t is the time in years when the future cost is incurred. As an example, if the discount rate r is 12% and a sum of $ 10,000 is required after 5 years for maintenance or renewal of a particular system, the present value of the sum will be:

$$PV = \frac{10,000}{(1.12)^5} \text{ or about } \$5,675$$

The present value of all maintenance costs to be incurred at intervals of say 5 years up to the life of the structure can be calculated in a similar manner and added to the initial investment to give the net present value (NPV) of a particular system. Similar values for other protection systems may also be calculated and the results may be compared for financial evaluation.

Essentially Life Cycle Costing of a protection system involves four parameters, viz. initial costs, future costs, life of the protection system and discount rate. Initial costs of any protection system can be obtained readily from the supplier of a system. Future costs are the periodic minor maintenance costs, if any, plus the renewal costs at the expiry of the life of the system. It is, however, not easy to forecast these costs over a long period, as these are dependent on uncertain future economic scenario such as inflation, demand of the product, change in the costs of labour and other associated inputs, etc. The life expectancy of a protection system is equally difficult to predict, as this depends not only on the intrinsic protective quality of the materials used but also on a range of variable factors, such as workmanship and supervision. Thus, a protective system applied under close supervision, on properly cleaned surface, by experienced workmen with a particular flair for quality may last for say 8 to 9 years. However, the same system, if applied on an unclean surface, under inadequate supervision, by inexperienced workmen may not last even half that time.

The life of the protective system may also vary with the change of environment, such as sudden industrial growth in the area causing aggressive pollution effects for which the system was not initially designed. Additionally, calculation of the discount rate, though apparently simple, depends on several uncertain parameters such as inflation, interest rates, tax legislation, economic policy of the Government, etc. It is thus not easy to predict the rate with certainty. However, for comparing different competing systems, the same discount rate can be applied for each system to give sufficiently accurate results for ranking purpose. In the USA the discount rate is taken at 10%. In the UK the rate used by the Department of Transport is 8% without allowing for inflation. In Germany, it is 3% and is linked to national growth rate. In general, in developed countries the rate varies between 6 and 8%.

The above uncertainties are real, but are so in varying degrees only. In LCC computation, these three uncertain parameters, viz. future costs, time and discount rate are subjected to sensitivity tests for more accurate results. In order to assess the effects of the variability of the parameters, a number of computations are carried out by varying the parameter(s) related to maintenance and arrive at a set of present value costs from which the final choice is made by sound engineering judgement and other decision criteria. No doubt, a certain amount of inaccuracy may creep into the estimates. Nevertheless, the results will at least give the authorities an idea about the impact of future expenses even if these may not be absolutely correct. In any case LCC approach provides a better estimate than many other approaches and therefore can be utilised as a very useful qualitative tool for decision making.

BIBLIOGRAPHY

1. *Bridge Inspection Guide*, Her Majesty' Stationery Office, London, 1983.
2. *Manual for Maintenance Inspection of Bridges*, American Association of State Highway and Transportation Officials (AASHTO), Washington D.C., USA, 1974.
3. Chandler, KA and Bayliss, DA: *Corrosion Protection of Steel Structures*, Elsivier Applied Science Publishers, Barking, England, 1985.

Index